Henry E Colton

Coal

Coal mines of Tennessee and other minerals

Henry E Colton

Coal
Coal mines of Tennessee and other minerals

ISBN/EAN: 9783744790529

Printed in Europe, USA, Canada, Australia, Japan

Cover: Foto ©berggeist007 / pixelio.de

More available books at **www.hansebooks.com**

REPORT OF
HENRY E. COLTON,
GEOLOGIST and INSPECTOR OF MINES.

—ON THE—

COAL MINES OF TENNESSEE,

AND OTHER MINERALS,

—TO—

A. W. HAWKINS, COMMISSIONER,
AGRICULTURE, STATISTICS and MINES.

NASHVILLE:
ALBERT B. TAVEL, PRINTER TO THE STATE,
1883.

BUREAU OF AGRICULTURE, STATISTICS AND MINES,
NASHVILLE, March 23, 1883.

To his Excellency, W. B. Bate, Governor of Tennessee:

In my biennial report to Governor Hawkins, under date of January 1, 1883, in speaking of the coal mines of Tennessee, I used the following language (page 14):

"Coal mining in Tennessee has reached a condition which imperatively demands that the strong arm of the law should intervene to protect the miners from such catastrophes as have too often occurred, when, either from ignorance or criminal carelessness on the part of those in control, the necessary precautions for safety had been neglected.

"A supervision of these mines by a competent engineer is imperatively demanded. In order to meet what I regarded as a pressing public necessity, I directed Prof. Colton to make an inspection of these mines, a report of which will be submitted in a few days."

In reference to the iron interests of our State, in the same report (page 15-16), I said:

"The iron interests of Tennessee are developing into immense proportions, both present and prospective. Those who assume that the iron interests of our State are fully known, but advertise their own ignorance of the subject. Much, it is true, has heretofore been known and published in regard to our deposits of iron ores. Concerning many of them, however, the public have had no definite information in reference either to their extent or quality. With the view of securing such additional knowledge of our iron regions as may lead to further enterprises for the development of their riches, I have personally, and by the able assistance of Prof. Henry E. Colton, had surveys of different localities, the results of which are given in his report herewith submitted. A more detailed report of his survey of the Western Iron Belt will be submitted as soon as the necessary maps and drawings can be prepared, and will add much to the public information on this very important iron region of our State."

In compliance with the promises contained in the foregoing extracts, the following report by Prof. Colton is respectfully submitted.

A. W. HAWKINS, Commissioner.

N. B.—When this book was first projected, it was intended only to contain Prof. Colton's "Report on the Coal Mines," but the great demand for a book descriptive of our coal fields has caused him to be directed to elaborate it into such a work.

The Tennessee Coal Field.

The superficial area in the State of Tennessee, covered by coal bearing strata, amounts to 5,100 square miles, but this does not fairly represent the amount of coal in the State, as all this area has at least one seam of coal, a large proportion more than two, and a very considerable area has six or more workable seams. It is thus seen that the mere area of a coal field may be a very poor indication of the quantity of coal it contains, and without an examination into the thickness of the seams, and the quality of the coal therein, any judgment formed from area alone may be very incorrect. Missouri contains vastly more coal area than Tennessee, yet one seam in Tennessee is worth more for economic purposes than all the coal of Missouri.

In Pennsylvania there is a formation under the regular coal series, sometimes called the False Coal Measures, but classed by Prof. H. D. Rogers, as the Umbral Series, having only thin bands of coal; in Tennessee, these measures contain several workable seams of coal of excellent quality. The Lower and Upper Measures of Pennsylvania also appear in this State, but the great mass of rocks of the barren measures appear in much reduced thickness. It is thus seen that while Tennessee has all the bituminous coals of Pennsylvania, this State has also a coal-bearing strata, which in that is usually bare of any productive seams. While the area covered by our coal field is not so large, yet it is probable that we have as much of this mineral fuel—the anthracite field excepted—as the great iron State.

The Tennessee coal field belongs to that division known in geology as the Appalachian coal field, which, commencing in Pennsylvania, extends over a part of Ohio, Kentucky, West Virginia, Tennessee, and ends in Alabama. While its width in Pennsylvania and Ohio extends through nearly four degrees of longitude, at the northern boundary of Tennessee it is only about seventy-one miles wide, and at its southern boundary thirty miles. In its southern course into Alabama, it expands into a heart-shaped area one hundred miles or more in width. The area of this coal field in Tennessee includes within its limits the counties of Scott, Morgan, Cumberland, the greater parts of Fentress, Van Buren, Bledsoe, Grundy, Sequatchie and Marion, considerable parts of Claiborne, Campbell, Anderson, Rhea, Roane, Overton, Hamilton, Putnam, White and Franklin, and small portions of Warren and Coffee.

A part of this area possesses a peculiar topographical feature which has given it the name of the Cumberland Table Land, which name has been indiscriminately applied to the whole coal field, while among those who live there and have studied these features the terms Cumberland Mountains and Cumberland Plateau are distinctively used; there being no plateau character to the northeastern section, an area of about 1,000 square miles. This area is a series of short irregular mountain chains, breaking off from the main mountain, which is the divide between the waters of the Cumberland and the Tennessee, and frequently having peaks of great height. The area of the true table land or plateau country, is chiefly the northwestern part of the coal field, though the plateau character extends in some degree to the southwestern region, and even on to that part of mountain west of Chattanooga, improperly called Walden's Ridge. The table-land and plateau character is again found in Sand Mountain of Alabama. Still further south in that State, the coal measure rocks sink below the water level, and the superincumbent strata make a country of merely gentle rolling ridges.

The peculiar topography of this region may be understood by the following record of elevations, copied from Maj. Falconet's survey of the Tennessee & Pacific railroad. This survey ran an almost east and west line from Nashville to Knoxville, and crosses diagonally the northeast and southwest direction of the coal regions. The distances are not here noted, but on top of the table land, the second bench, the top of the great conglomerate sandstone, for twenty-six miles, there is only a variation from a level of 142 feet:

N. & C. R. R., Nashville above tide water	476
Another survey gives the elevation of Nashville as 435 feet above the sea	
Lebanon	559
Hawkins' Gap Summit	704
Round Lick Ridge	662
Gardensville	501
Pea Ridge Summit	1,079
Allison's	1,110
First bench of Cumberland Mountain	1,477
Second bench of Cumberland Mountain	1,831
Standing Stone	1,876
Summit of Plateau	1,918
Highest point of Survey	1,973
Bledsoe's Stand	1,880
Langley's x Roads	1,347
Big Emery river, near Montgomery	1,039
Wartburg	1,379
Winter's Gap	834

It is seen from this that the rise from the valley at Allison's to the first bench is rapid, and thence to the second bench still greater. The distance from Allison's to second bench is only twelve miles, while the rise is 720 feet. The fall from the plateau at Bledsoe's stand to the region of the influence of the great down throw at Langleys, and on to the river at Montgomery is equally wonderful.

The most valuable series of elevations yet made through this region, were for the Cincinnati Southern railroad, and through the kindness of the officials of that road, I have been furnished with them and much other valuable matter. It is worthy of remark that they give the same elevation to Bledsoe's stand as that given by Maj. Falconet. The Trustees of that railway had surveys made from Columbia Ky., through Clay, Overton, Putnam, White, Van Buren and Sequatchie counties to Chattanooga; also from Monticello, Ky., through Fentress, Cumberland, Bledsoe and Sequatchie; also from a point on the last line near Crossville to White's creek bridge in the valley, where the road has since been built.

The first line followed the valley at the Northwestern base of the Cumberland mountain to Sparta, and eleven miles east of that town, commenced ascending the mountain and crossed over to Sequatchie valley. The elevation of Sparta is about 950 feet.

[Since the above was put in type Mr. Hunter McDonald, of the Nashville & Chattanooga Engineer Corps, has sent me his levels, which, connected with Col. Jones', makes Sparta 976 feet above the sea.]

As stated the ascent of the mountain proper commences in a valley eleven miles east of Sparta 860 feet above the sea, thence in about ten miles the rise is 837 feet to an elevation of 1,700 feet above the sea; thence for eleven miles the rise is to a summit of 2,000 feet; thence to Dunlap the fall is 1,250 feet in sixteen miles. Hence the part that may be called the plateau, has a variation of 200 feet of elevation in eleven miles; while further north we have seen there is a variation of only 142 feet in twenty-six miles.

The second line of the Cincinnati Southern Railway surveys entered the State north of Jamestown. At this point the valley where the rise commences is in Kentucky, fifteen miles north of the State line. The elevation there is 900 feet, at the State line it has reached 1,650, and from thence to Bledsoe's Stand there is a rise of only 230 feet, the distance being fifty miles; thence to Obed River there is a rise to a summit of 1,900 feet between Obed River and Clear Creek, another similar elevation between Obed River and Daddy's Creek, another, but more sharply defined, at Crab Orchard, and then the mountain falls rapidly to the Tennessee Valley on White's Creek. From the point where Crab Orchard is crossed to the Valley is fourteen miles, and the total descent 936 feet.

The other line was run down to the west of the head of Sequatchie Valley. It adds nothing of special interest except showing the irregular escarpment-like form which the rim of the valley there assumes, being raised higher than the mountain behind it. The descent to Pikeville, in nine and one-half miles, is 1,050 feet. Pikeville is about 850 feet above the sea level, and from Pikeville to Dunlap, twenty miles, this singular valley falls only one hundred feet.

The Cincinnati Southern Railway was constructed on a line entirely different from any of these. It crosses the mountain nearly on the boundary line of the action of the great downthrow. Leaving Chattanooga at an elevation of 635 feet above the sea, it follows the Tennessee Valley to Emery Gap for near eighty miles, and has there only attained an elevation of 792 feet, eighteen miles beyond, near Wartburg, where Emery River is left, at Triplett's Gap it has reached 1,209, and ten miles further on, where is the water divide, the grade elevation is 1,430, and that of the top of the mountain 1,585. Between New River and the State line a narrow strip of the plateau is passed over, the elevation for ten miles ranging from 1,598 to 1,560. From this there is a fall at Chitwoods, the State line, to 1,320; beyond this is another arm of the plateau where there is an elevation for nine miles from 1,400 to 1,450, and for twenty-five miles to the first bench of the Cumberland Mountain, placed at 1,257, then slowly down to 700 feet, the elevation of the Cumberland River. The level of the rail on the bridge at Cincinnati is 537 feet above sea level, 98 feet lower than the terminus in Chattanooga.

The ascents of these railroad lines do not properly represent the wall-like character of the outer edge of this singular region. There are numerous places on the western boundary, where in two hundred yards of horizontal level the strata drop a thousand feet in vertical line, from the upper coal measures down to the lowest subcarboniferous limestones. There are equally as many points on the eastern boundary where a horizontal level of five hundred or less feet will represent a vertical fall of from 1,000 to 1,500 feet, and a change from among the upper carboniferous rocks down into the Niagara limestone and the Clinton iron ores.

The part of our coal field area therefore which should be properly called the Cumberland Table Land or Plateau is an area extending from the Kentucky line in Scott and Fentress counties, southwest to Franklin, and narrowing in width as it passes southward.

The main floor on top of this table land is a sandstone, sometimes a conglomerate, which is met everywhere when one reaches the main bench of the mountain. It runs back on the mountain for varied distances, rising upon it at these distances are ridges from fifty to a hundred feet in height, the tops of which are more or less capped with sandstone, and form plateau areas of greater or less width.

This is a peculiar feature in the southwestern section, and the manner in which these ridges are built up on the conglomerate is excellently illustrated at Tracy City; but to the north, as the Kentucky line is approached these ridges cease, and the conglomerate forms the floor of a vast area of almost level country, and on the western edges of the table-land is a steep escarpment or brow, bold, distinct, and well marked from twenty to one hundred, and sometimes two hundred feet high. Beneath this often overhanging brow, the gradual slopes of the sides begin and run down to the low lands. Just below the lower coal measures is usually a character of flat, called the bench of the mountain, heavily timbered with walnut, poplar, buckeye and beech, its floor being the upper mountain limestone; from it commence the slopes which are composed of the various sub-carboniferous limestones and their accompaning shales The western edge is jagged, notched by innumerable coves and valleys, and presenting a scolloped or ragged contour, with outlying knobs separated from the main Table-land by deep ravines or fissures. The eastern outline of the Cumberland Mountains is, for some distance, a'nearly direct line, making, however, a curve in Anderson and Campbell counties. In the southern portion, near the eastern side, is a deep gorge, canoe-shaped, with steep escarpments rising eight hundred to one thousand feet above the valley, through which the Sequatchie river flows. This is the Sequatchie valley which separates the lower end of the Table-land into two distinct arms. Through the eastern arm the Tennessee river breaks, and after flowing down the valley for a distance of sixty miles, turns at Guntersville, Alabama, and cuts through what is there left of the western arm fifty miles from the Tennessee line. This Sequatchie Trough is one hundred and sixty miles in length, the Tennessee part being sixty miles, and that in Alabama one hundred.

The southeastern arm of the coal-field, on the western side of which is the Sequatchie Valley, is eight miles wide. Between the Tennessee river and the Nashville and Chattanooga railroad, it is called Raccoon mountain. Separated from this latter by Will's valley, is Lookout mountain, an outlier of the Cumberland Tableland it may be called, probably at one time connected with it; of the same geological formation, but not containing any workable coal at any point yet discovered in Tennessee.

(11)

The position of the coals and other strata in this northwestern division of the Tennessee coal field may be gathered from a study of the following section made near Tracy City by Dr. Safford, being made from the mountain limestone up the gulch of Fiery Gizzard, and on reaching the conglomerate plateau to that place and the ridge, or second bench, in which the Sewanee seam of coal is located.

First in ascending order is the mountain limestone and its shales, hundreds of feet thick. Upon it are:

FEET.
1. Shale and thin sandstone................20
2. Hard sandstone................20
3. COAL—No. 1, sometimes a thin shale above and below it—coal................1 to 3
4. Sandstone, hard................78
5. Sandy shale................22
6. Shale with few inches of clay................8
7. COAL—No. 2, outcrop................½ to 1½
8. Sandstone (cliff rock)................65
9. Shale with clay at top................10
10. COAL—No. 3, outcrop................½ to 1
11. Conglomerate................70
12. Sandstone................17
13. Shale................3
14. COAL—Outcrop................1
15. Shale, some sandy................45
16. COAL—(the Sewanee seam)................2 to 7
17. Shale, sandy................45
18. Sandstone................86
19. Sandy shale................25
20. Dark clay shale................1
21. COAL—outcrop................½
22. Shale................23
23. COAL, only a few inches thick................
24. Conglomerate, (usually a sandstone, and when the ridges are broad enough to make a plateau, is then floor rock)................50

This section reaches to the top of the highest ridge above the Sewanee seem, and is stated to be 2,162 feet above the sea, and the elevation of the Sewanee seam at its outcrop is 1,922 feet above the sea. As seen from the elevations given heretofore, the elevations to the north are less, but at the same time it is known that for some distance in that direction this seam is found, though undoubtedly with less covering of sandstones and shales.

Another section was made at Bon Air, and is as follows:

MOUNTAIN LIMESTONE.

		FEET.
1.	Sandstones and shales	20
2.	Fine clay	1 to 2
3.	Coal—2 to 4 feet, average	3 to 2½
4.	Shale	12
5.	Coal—thin seam and fire clay	1 to 2
6.	Sandstone	13
7.	Shale—probably coal has iron stones	80
8.	Conglomerate.	

This is a section of Little's Bank, two and a half miles from Bon Air proper and six miles from Sparta.

The above section was taken by Dr. Safford in 1867. Since that time the bank has been more extensively opened, and the lower coal, as measured during my late visit to that region, was very uniformly 3 to 3½ feet in thickness. The middle seam (5 of the section) had also developed to a coal never under 2½ feet thick. The upper seam, immediately under the conglomerate, was 14 to 16 inches thick, and is undoubtedly the same as the Cliff seam of the Ætna mines. Col. Samuel S. Jones, of the Nashville, Chattanooga & St. Louis Engineer Corps, ran a line from the depot at Sparta to the Bon Air Coal Company's property, terminating at what is known as the Fitzwater mine. This survey determined the elevation of that seam as 1,585 feet above the sea level. It is a singular fact that the same seam in Poplar Mountain (Kentucky) by the United States Engineer's survey, is 1,436 feet above the sea level, these coal measures being lower there, as the incline to go beneath the Upper Measures of East Kentucky has there commenced. In the great rock which makes the Bluff at Bon Air, which is a prominent point from all points of the Sparta Valley, I found cut by the Engineer Corps of the Tennessee & Pacific Railroad an elevation above the sea of 1,827 feet. It must be remembered that the Nashville, Chattanooga & St. Louis Engineer's elevation is from a base at Nashville of 435 feet, while that of the Tenessee & Pacific is from a base of 476. Not an important matter, however, in the point being considered, which is to illustrate the great dip of strata to the southeast. A direct southeast line would strike near Spring City, in Rhea County, there the mountain limestone is low down in the valley. It is true, however, that the dip to that point is not regular, as the action of the Sequatchie Valley fold is intermediate, and there is a synclinal between Crab Orchard Gap and the mountain rim near Sparta, to which the strata dip on both sides, the more severely from the eastern or Crab Orchard side, as the elevation in the Gap is about 1,700 feet, the mountain limestone showing in the road at the Gap. In the intermediate strata as they rise, all the coals show; in the higher ridges the Sewanee seam, the drop on that side seeming to have preserved it, as it is gone for some distance from the rim on the Western side.

The chief seam of coal of the area which we have discussed as the northwest division of the Tennessee coal-field, is known as the Sewanee seam, from the fact that it has been the only one worked with success and to any extent for a length of time on the Sewanee Mountain at Tracy City. It extends over a greater area than any other seam of coal in the State; at the same time the sub-conglomerate measures afford in White, Putnam, Overton, Fentress, and in Kentucky, seams of remarkable thickness, having no superior anywhere for domestic and steam purposes.

The usual classification of the coal seams of the Tennessee coal-field has been into upper and lower measures, the division being made on the thick conglomerate which forms the cap of the plateau. The proper division, however, is into Lower, Middle and Upper measures. The Lower Measures representing the sub-conglomerate strata; the Middle those of the Sewanee section (Nos. 12 to 24 inclusive), and the Upper the coals appearing above water level in the horizontal strata of the high mountains in the Eastern division of the coal field. The measures thus classed as Lower Measures are sometimes called False Coal Measures in Pennsylvania, and those we have denominated Middle Measures are there called the Lower Measures. . A more proper classification and one more in unison with those of other States is into Upper, Lower and Sub-conglomerate measures. The classing of our sub-conglomerate measures as lower and the Sewanee section as upper, tends at least to produce confusion, if it is not absolutely inadmissible, when any comparison of our coals with those of Pennsylvania is attempted, the fossils of the Sewanee seam being identical with those of the Lower coal measures of Ohio and Pennsylvania, which do not exist in the horizontal seams of Coal Creek; while it is also evident that the eighteen coal seams existing above water level at that place and at Poplar Creek do not belong to the Lower measures. Our Upper measures certainly correspond in some degree to those similarly termed in West Virginia and Pennsylvania, but the constituents of our strata are so different, also the conditions under which our coals were formed, as well as the disturbances to which they were afterwards subjected, that any satisfactory comparison is at least difficult.

The Sewanee seam furnishes a larger amount of coal than any other single seam in Tennessee, and has all the qualities that combine to make a useful and valuable coal. It varies in some of its characteristics and constituents in different localities, but that is a common freak of all coal seams in every coal-field. It makes a good coke, is a good steam-making coal, makes a hot, durable fire in the grate, and is nearly free from sulphur. It is found in the elevated ridges all over the Table-land, but in the horizontal strata of the Coal Creek and Winter's Gap section of the field it has sunk far beneath the surface. It is the main seam of the pitched strata of Walden's Ridge, and continues therein with much persistency from Rockwood to near Careyville, and from beyond that place to near Cumberland gap. Where the Ridge is regular in surface, and the strata in place, the seam is of regular thickness and easily worked, with a certainty of obtaining a constant supply, but where the strata are broken by ravines or gorges, it is also disturbed, sometimes lost entirely, and again rising into great thickness. If it continues under the horizontal strata in the region northeast of Dayton, as there is every reason to believe, the area in which it may be made available in the present or future will not be much, if any, less than 3,000 square miles, which, at the low estimate of 3,000 tons to the acre, would yield 5,760,000,000 tons of coal, the digging of which would keep 1,000 men employed 1,440,000 days, or 4,800 years, if they worked 300 days per year and dug four tons per day; and their wages at two cents per bushel would amount to $600,000 per year, or for the whole time, $2,880,000,000, the labor value of one seam alone of all the Tennessee coal field.

The other grand division of the Tennessee coal field lies to the east of the Cincinnati Southern railway, is irregular in shape, and its topography is as wild and varied as its shape is irregular. It is simply a mass of mountains, all, however, as heretofore said, having a connecting link to the main range, which is the water divide. Properly, this area should be bounded by the Sequatchie Valley and a line from its head to the Cincinnati Southern road; thence with that road for a short distance, and then a line bearing a little north-northeast from it. This is the area on which the influence of the Sequatchie and Pine Mountain fault was felt and the great down-throw acted, whereby the strata we have previously considered and shown as existing on top of mountains over a thousand feet above the sea level have, on the eastern side of this coal field, been thrown down a thousand feet below the surface, and also at another point changed in their physical position so as to be vertical instead of horizontal. On the top of these sunken strata are high ridges and mountains, containing coal seams unknown to the area we have previously considered. These seams and the strata accompanying them are approximately horizontal, but immediately on their eastern extreme rises a peculiar wall-like ridge in which the strata are pitched at an angle varying from twenty degrees at Cumberland Gap to sixty at Big Creek and Coal Creek, to thirty at Rockwood, and losing its distinctiveness as an independent ridge south of Dayton, has its strata in the mountain nearly horizontal. This ridge of upturned rocks is called Walden's Ridge, a name improperly applied to the arm of the mountain between Sequatchie Valley and the Tennessee Valley, which is eight miles wide, and in which the strata are horizontal.

This Walden's Ridge is an outlier of the Cumberland Mountain, a vast wall of upturned rocks, ranging from six hundred to twelve hundred feet high. This singular formation is best seen north of Big Emery Gap. A base line drawn horizontally through the ridge at that Gap or at Coal Creek would probably give a width of twelve hundred feet. The line of demarcation between the inclined strata of Walden's ridge and the horizontal layers of the Cumberland mountains is sharp and well defined. Within a few feet one steps from the almost vertical sandstones of Walden's ridge to those of the Cumberland mountain lying horizontal. Behind he sees the steep inclined crags of the Ridge, and in front the shales, slates, and sandstones lying one on the other. This ridge is most continuous and conspicuous in its tilted strata from Big Emery Gap to near Careyville, but those peculiar characteristics are gradually lessened to the southwest from Emery Gap until near Dayton the dip of the strata is very slight, and as stated the distinction between the two is lost. The greatest action of the down-throw, therefore, took place between Emery Gap and Careyville, and again from Careyville to Cumberland Gap. To its action, says Prof. Lesley, is due the preservation of the numerous beds of coal in the high mountains on Poplar Creek, at Winter's Gap, on Coal Creek and at Big Creek Gap.

The peculiar relation of these strata to each other is best illustrated by the following diagram:

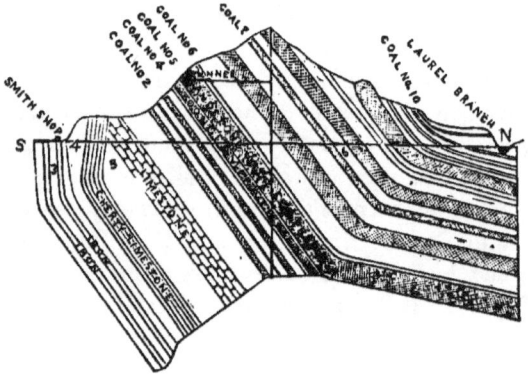

The accompanying cut will give an idea of the peculiar topography of Tennessee, and the relation which the higher coal measures bear to the East Tennessee Valley and the Middle Tennessee Basin, as well as the faults which have brought the varied strata to the surface on the eastern side of the Cumberland Mountain. It would be almost impossible, in so small a space, to show fully the peculiar uplift of the Walden's Ridge.

Further, the cut shows the mountain limestone too, far up the mountain on the eastern side, and does not make the coal measures there thick enough in proportion to the thickness given to the mountain limestone and the siliceous formation. The Clinton at the eastern base of the mountain, though a very important formation as carrying the valuable red fossil iron ore, is omitted, because very narrow. For the same reason the Devonian is also left out, because it has only one member represented, the black shale, which though very persistent, is nowhere over 75 feet thick, and reaches that only at Cumberland Gap.

This cut very fairly represents a line across the State from Hickman, Ky., through Nashville, Crossville, Kingston and Maryville, to the North Carolina line. Farther south would strike the Sequatchie Valley, farther north the Irregular Mountains.

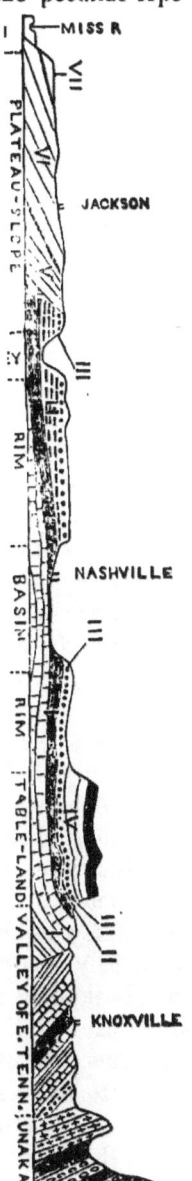

I. Is the Knox Dolomite.
II. Nashville and Trenton.
III. Siliceous.
IV. Mountain Limestone.
V. Cretaceous.
VI. Tertiary and Quarternary.
VII. Alluvium.

The following sections taken by Prof. Bradley near Little Emery Gap and at Coal Creek will illustrate the position of the strata. Commencing on the western edge of the valley we have:

1. Red, greenish and yellowish shales, including two bands of red fossil iron ore......103 to 150 ft. thick.
2. Black shale.. 90 "
3. Black and drab shales........................ 26 to 117 "
4. Green and drab shales........................ 3 to 24 "
5. Limestone with heavy bands of chert...... 160 "
6. Shale or shaly limestone...................... 125 "
7. Blueish drab fossil limestone................ 35 to 200 "
8. Shales with Coal No. 1....................150 to 200 "
9. Thin bedded sandstone........................ 45 to 50 "
10. Dark, drab compact sandstone............... 40 to 50 "
11. Gray, ferruginous shales with Coal No. 2. 170 "
12. Heavy bedded fine sandstone................ 25 "
13. Clay shales, part sandy, with Coal No. 3.. 180 "
14. Heavy bedded sandstone, mostly conglomeric just above Coal No. 4..................140 to 150 "
15. Hard drab shales............................... 2 to 3 "
16. Coal No. 5. .... 3 to 4 "
17. Ferny shales, some sandy..................... 40 to 4 "
18. Heavy bedded fine sandstone................ 31 to 4 "
19. Soft clay shales................................ 2 to 4 "
20. Heavy bedded, ferruginous sandstone, light colored, part pebbly................. 5 "
21. Sandy shales.................................... 8 "
22. Heavy bedded, coarse and fine sandstone.. 53 "
23. Shales.. 8 "
24. Heavy bedded sandstone..................... 47 "
25. Shaly sandstone............................... 10 to 15 "
26. Dark, drab clay shale, with fine clay....... 5 to 6 "
27. Coal No. 6...................................... 3 to 6 "
   This seam of coal Prof. Bradley believes to be the equivalent of the seam worked at Rockwood and at Tracy City, and he is undoubtedly correct.
28. Dark drab to black and ferruginous clay shales.................................... 25 to 30 "

29.	Shaly sandstone.....................................	12 to	16	ft. thick.
30.	Shales...	70 to	80	"
31.	Heavy bedded sandstone........................	40 to	45	"
32.	Sandy shale and ironstones, probably Coal No. 7...	180 to	200	"
33.	Irregularly bedded sandstone.................	50 to	70	"
34.	Ferruginous shales with beds of iron stone.	90 to	100	"
35.	Shales, shaly sandstones, thin................		225	"
36.	Heavy bedded sandstone........................		153	"
36.	Sandstones, shales, sandy and thin.........		193	"

This ends the vertical rocks, and up the mountain, across the little stream which divides it and the ridge, about 700 feet of shales and sandstones are noted, containing five seams of coal. The ridge at this point is higher and the mountain not so high as at Coal Creek. At the latter place Prof. Bradley made the following section :

COMMENCING IN THE EDGE OF THE VALLEY :

1.	Limestone, part cherty............................	100 to	300	feet
2.	Sandstones and shales (coal 1).................	200 to	300	"
3.	Heavy bedded sandstones........................	50 to	60	"
4.	Shales and sandstones, thin laid...............		150	"
5.	Coal No. 2..	1½ to	2	"
6.	Dark compact clog shales	55 to	60	"
7.	Heavy bedded sandstones........................	30 to	35	"
8.	Shales and sandstones...............................		55	"
9.	Coal No. 3..	3 to	4	"
10.	Shales and sandstones...............................	40 to	50	"
11.	Hard dark shale......................................	15 to	20	"
	About this point is the change from vertical to horizontal rocks, the creek runs on the line of change.			
12.	Underclog	2 to	4	"
13.	Coal (No. 4, first of the horizontal)..........	1½ to	2½	"
14.	Clay shales, flag-stones, sandstones...........		35	"
15.	Drab to black clog shale with ironstone.....		30	',
16.	Shales and sandstones...............................	30 to	40	"
17.	Underclay...	1 to	2	"
18.	Coal No. 5–2 of horizontal seams...............	4 to	6	"

19.	Claystone		15	feet.
20.	Clay and coal mixed		1	"
21.	Coal (6–3)	1½ to	2½	"
22.	Laminated sandstone	12 to	15	"
23.	Clay shales with iron-stone	130 to	150	"
24.	Sandstones		40	"
25.	Coal (7–4 of horizontal)	2 to	3	"
26.	Black slaty shale		2	"
27.	Sandstones and shales		140	"
28.	Black slaty shale			"
29.	Coal (8–5 of horizontal)		2	"
30.	Shales and sandsones		190	"
31.	Coal (9–6)		?	"
32.	Black bituminous shales		10	"
33.	Shales and sandstones		290	"
34.	Laminated sandstones		30	"
35.	Coal (10–7)		3½	"
36.	Shales and heavy clifty sandstones		180	"
37.	Coal (11–8)		2½	"
38.	Shale		10	"
39.	Coal (12–9)		3	"
40.	Shales and sandstones		100	"
41.	Coal (13–10)		?	"
42.	Shales and sandstones		110	"
43.	Coal (14–11)		?	"
44.	Shales with thin sandstones		350	"
45.	Coal (15–12)	5 to	7	"
46.	Shales with some heavy sandstones		309	"
47.	Coal (16–13)		2½	"
48.	Shales and heavy sandstones		20	"
49.	Coal (17–14)		1½	"
50.	Shales and sandstones		10	"
51.	Coal (18–15)		1	"
52.	Shales with irregular ironstones		20	"
53.	Coal (19–16)		3½	"
54.	Udner-clay and sandy shales		20	"
55.	Coal (20–17)		1½	"
56.	Shales and heavy sandstones		80	"
57.	Coal (21–18)		½	"
58.	Shales and sandstones		200	"

From the above it is seen that there are at that place eighteen seams of coal in the horizontal strata above water level, and that seven of these are over two and a half feet thick, making a total of thirty-one feet of workable coal, with a possibility that beyond the outcrops, the others also exceed that thickness.

It has been assumed that the vertical seams of Walden's Ridge go down under and become horizontal. From the slope and drift now being made at Rockwood, this appears to be true. That there is below water level a great thickness of coal measures is proven by the boring of the salt well, made by E. A. Read, at Winter's Gap. The well was sunk just west of the junction of the horizontal and vertical rocks, in the former. A record of the boring is as follows :

SURFACE.

1 to	8 feet.		Mixed sand and clay.
8 to	10	"	Quicksand.
10 to	12	"	Slate.
11 to	12	"	Fire clay.
12 to	14	"	Very good coal.
to	40	"	Mostly slate rock, with seams of hard shaly rock, and thin seams of coal.
40 to	43	"	Very hard rock.
40 to	43	"	Black gritty slate.
44 to	73	"	Mostly slate.
73 to	75	"	Seam of coal.
75 to	90	"	Slate rock.
90 to	91	"	A crevice—tools dropped.
A 91 to	137	"	Quite hard slate, gritty seams in it.
137 to	139	"	Seam of soft coal.
139 to	147	"	Mixed fire clay and seams of coal.
147 to	160	"	Slate and some small seams of soft coal.
B 160 to	282	"	This rock is very hard, most of the 122 feet particles of mica in it, a few pieces coarse and soft, and very white.
282 to	283	"	Small seam of slate.
283 to	330	"	Sand rock, some fine white seams.
A 330 to	383	"	Mostly black slate, some seams of fine clay then hard shale—the sand pumped out soon turned rusty. Struck a little salt-water river at 383

383 to 399 " Found here the hardest rock we ever struck.
389 to 427 " Very hard.
427 to 431 " Black rock, water oily.
431 to 464 " Dark slate.
464 to 490 " Good sand rock to drill.
490 to 514 " Hard slate.
514 to 523 " Rock resembling soapstone very much.
523 to 560 " Very hard rock, white when dry.
560 to 584 " Hard rock, salty to the taste, and turned rusty on exposure.
584 to 588 " A crevice of about 15 inches with sweet water.
588 to 600 " Hard rock.

At this point the drilling was stopped.

A well had previously been bored at this place by Prof. Estabrook, a gentleman of high scientific attainments and much energy. Mr. E. F. Wiley, who assisted him in the work, informs me that one of the seams of coal they passed through was three feet thick. He did not remember the exact depth below the surface. The water from both these wells was weak, not over six degrees. Prof. Estabrook, at considerable expense, erected high walls of briers to condense the water, and made considerable salt, until his briers were burned by some vandals. Mr. Read expended a great deal of money, but from defective tubing the surface water was not kept out, and in view of the cheapness of manufacturing salt in Virginia and on the Kanawha, combined with cheap transportation, it was not deemed profitable to go into the manufacture on an extensive scale. During an examination of Mr. Read's operations for some Knoxville gentlemen, I made salt from the water he was pumping out at the rate of one bushel to two hundred gallons of water.

On the western side of the coal field the general dip of the strata is slightly to the northeast. The elevation of the sub-carboniferous limestone on the mountain side near Tracy City is about sixteen hundred feet above the sea. On a direct east line, near the foot of Walden's Ridge, the same rock is only about seven hundred feet above the sea; on the line of the Tennessee & Pacific road, in Putnam county, the limestone is about fourteen hundred feet above the sea, while in a direct east line, near Winter's Gap, in the valley where its position is nearly vertical, and it is supposed to dip under the mountain with the other inclined strata, its outcrop is only eight hundred feet above sea level. The level of the valley at Cowan is nine hundred and seventy-three feet above sea level, and the level of the Sewanee seam at Tracy City is nine hundred and forty-nine feet higher. This seam dips to the southeast about eight feet to the mile; hence from a location in Fentress, in the fifty miles distance to Winter's Gap, it would be deep down under the horizontal strata of the high mountains, though coming up again above the valley in the inclined strata of Walden's ridge.

As stated, at Tracy City this seam is 1,952 feet above the sea, while at the Soddy Mines, very near a direct east line from Tracy, the elevation of the same seam is 1,221 feet above the sea level; the seam at both places being approximately horizontal, but showing a fall of 701 feet in the thirty miles distance. But this is not practically accurate as a uniform descent, for Sequatchie Valley intervenes and the fault which made that valley has thrown the sub-carboniferous limestone high up on the side of the mountain on the eastern side of the valley, though not nearly as high as it is near Tracy, and from my observation on the mountain, I am satisfied that on the top of the so-called Walden's Ridge, near the Sequatchie Valley, the dip of the coal strata is for a short distance very rapid to the east. On the west side of the valley the limestone is much lower down the side of the mountain, and the coal dips slightly to the west. This feature is a prevailing one in the neighborhood of all faults. The coal seams at Coal Creek and on Poplar Creek both having a western dip for a short distance, then forming a basin followed by a gentle rise to the west; the former caused by the influence of the Walden's Ridge fault, the latter by the great Sequatchie Valley synclinal and the still greater Cincinnati axis.

Towering high above the valley, in Anderson, Morgan and Campbell counties is the series of mountains heretofore mentioned. They reach an altitude of over three thousand five hundred feet above sea level, and contain coal seams to their very summits. Here is the equivalent of the Upper Measures of Pennsylvania. And it is safe to assume that the carboniferous strata in this region, estimating by the data derived from the boring of the salt well at Winter's Gap, attain a thickness of full four thousand feet in a direct vertical line from the top of the American Knob, or Brushy

Mountain to the lowest sub-conglomerate coal. At Careyville Dr. Safford determined the elevation of Cross Mountain, with nine seams of coal, to be three thousand three hundred and seventy feet above the sea, and two thousand three hundred and twenty-nine feet above the valley. This is at the northeastern end of the Upper Measures, as the still higher Brushy Mountain is near the southwestern end. In this distance of about forty miles, is the series of high ranges and peaks alluded to above. Hence we have in this district an area of about two thousand square miles, the greater portion of which contains, above water level, from four to seven seams of coal over three feet thick; thus showing, in this part of the Tennessee coal field alone an extent of thickness and a number of seams, available in the future, beyond the previous calculations of geologists.

For comparison a section of the Pennsylvania measures is given below as published in "Coal and its Topography," by Prof. J. P. Lesley, and here copied from Dana's Geology, commencing at the millstone grit and numbering up:

1. Coal [A] with four feet of shale .... 6 feet.
2. Shale and mud rock.............................. 40 "
3. Coal [B] equivalent of Mammoth............. 3 to 5 "
4. Shale with some sandstone and iron ore...... 20 to 40 "
5. Fossiliferous limestone........................... 10 to 40 "
6. Buhr stone and iron ore......................... 1 to 1½ "
7. Shale... 25 "
8. Coal [C] Kittaning and Peytona Cannel...... 3½ "

UPPER MEASURES.

1. Mahoning sandstone............................. 75 feet.
2. Coal [F]... 1 "
3. Shale, variable but considerable thickness......
4. Shaly sandstone................................... 30
5. Red and blue calcareus marls.................. 20 "
6. Coal [G]... 1 "
7. Limestone fossiliferous........................... 2 "
8. Slates and shales..................................100 "
9. Grey clayey sandstone........................... 70 "
10. Red marl .... 15 "
11. Shale and slaty sandstone....................... 10 "

12.	Limestone, not fossiliferous...............	3		feet.
13.	Shales..................................	32		"
14.	Limestone............................	3		"
15.	Red and yellow shale	12		"
16.	Limestone............................	4		"
17.	Shale and sand,.......................	30		"
18.	Iron ore spathic.......................	15		"
19.	Limestone............................	1 to	$1\frac{1}{2}$	"
20.	Pittsburg Coal [H].....................	8 to	9	"
21.	Shale, brown ferruginous sandy...........	30		"
22.	Sandstone, gray and slaty................	25		"
23.	Shale, yellow and brown................	20		"
24.	Limestone, (the meat includes two seams of Coal, 1 and $2\frac{1}{2}$ feet)...............	20		"
25.	Shale and sandstone....................	17		"
26.	Limestone............................	1		"
27.	Shale and sandstone....................	40		"
28.	Coal [K].............................	6		"
29.	Shale, brown and yellow................	10		"
30.	Sandstone, coarse brown................	35		"
31.	Shale................................	7		feet.
32.	Coal [K].............................	$1\frac{1}{2}$		"
33.	Limestone 4, Shale 4, Limestone 4, Shale 3....	15		"
34.	Shale 10, Sandstone 20, Shale 10..........	40		"
35.	Coal [M].............................	1		"
36.	Sandstone with 4 feet of shale...........	24		"

In the Tennessee coalfield the corresponding seams to this classification would be:

A. A small thin seam, found at Sewanee mines.
B. No. 6, at Emory and Rockwood, of the inclined strata; the main Soddy seam and the Sewanee seam of the horizontal strata.
C. The slate seam at Etna and Daisy, No. 2 of those worked at Soddy, not recognized elsewhere.
D. The Kelly coal of Etna and Daisy, and No. 3 of Soddy.
E. Walker coal of Etna and Daisy.
F. Beneath water level at Coal Creek.
G. No. 4, of Bradley's Coal Creek section.

H. The seam now worked at Coal Creek and Poplar Creek does exist in Tennessee south of Big Emery River, No. 5, of Bradley.

I. & J. } No doubt identical with 6 and 7 of Bradley.

K. L. & M. } Correspond in location to 8, 9 and 10, of Bradley; leaving 10, 11, 12, 13, 14, 15, 16, 19 and 20 existing above, and in addition to the seams of Pennsylvania.

Lesquereaux in his classification of the Kentucky coals, has four seams of coal between G and H, and classes the Pittsburg coal as No. 11 of the Kentucky series. The seam, however, in which he finds the supposed identity is in the Western Kentucky field. The peculiarity of the Pittsburg coal bed is an upper seam of somewhat shelly coal, in the upper part of which are some bands of sulphur, then a shale parting, then a body of excellent coal in lamina, then another shale parting, then a seam of hard firm coal. In the Connellsville region it is known that the middle bench makes the best coke. The same partings exist at Coal Creek and Poplar Creek, and in the tests for coke-making, it has been admitted that the middle bench there will make a coke equal to any in the world. By the classification and section made by Mr. Lesquereaux, No. 15 of Bradley's section would be the same as his Mulford coal and the Pittsburg seam. The matter is, however, of but little practical importance, as even if identical, it might not have the same good qualities, it being well known that coal seams will vary in their characteristics in a few miles; a notable instance of this is found in the Pennsylvania coalfield, even in the Pittsburg seam itself. At Valley Works the coal is of great excellence and purity, while at Latrobe, about twenty miles north, near the rim of the field, the coal is so impure that it is thought necessary that it should be washed to make good coke. Actual manufacture and use is the true test, but this will be discussed and analyses given in another part of this book.

As to the identity of the Sewanee seam there is more positive evidence. The roof is abundant in impressions of Lepidodendra, and this species is found only in the Lower Coal Measures, and, in fact, in those but sparsely above seam B. No traces of these plants is found in the roof of the seam worked at Coal Creek.

(27)

A very remarkable feature of the Tennessee coalfield is the utter absence of limestones, which are abundant in Western Pennsylvania. Though the geology of our coalfield has not been thoroughly investigated, sufficient is known to show that if having any at all, the limestones do not exist in our coalfield in the number or thickness found in Pennsylvania. In our coalfield, however, the clay iron stones or bands and nodules of carbonate of iron are very abundant, especially in the lower measures. The sub-carboniferous limestone exists in a very short distance of the coal seams both on the eastern and western sides.

REPORT ON THE
COAL MINES,
OF THE
State of Tennessee.

BY HENRY E. COLTON, INSPECTOR.

Dr. A. W. Hawkins, Commissioner of Agriculture, Statistics and Mines:

The following report is submitted for your consideration. I regret that time has not been allowed to make it more complete:

The coal area of the State is estimated as comprising 5,100 square miles. While some of the seams of coal in this area are spread over its whole extent, yet there are portions of it quite distinct in topographical features, as well as in the quantity and characteristics of the coal they contain. In other States such distinctions are classified under the name of districts. No such classification has ever been made in our State from the reason, probably, that the coal interest of the State is in its infancy, and has heretofore been confined to only a portion of the whole field. Such a classification is necessary, from the fact that the geological strata are different; that the topography of the respective sections is entirely dissimilar. The markets to which the coal is transported are seldom the same, and in great measure the coals are distinct in character and the uses to which they are applied. At the same time there exists in Ten-

nessee a wonderful freak of nature, by which coals, geologically
hundreds of feet lower, are brought up in close proximity to those
of a later age, and also adjoining formations created far anterior to
the carboniferous strata. This peculiar uplift of the lower carboniferous rocks is called the Walden's Ridge, and extends from Cumberland Gap to Dayton, in Rhea County. In it the coals are inclined at an angle ranging from 20° to 40°, and as the system of
mining in it is entirely different from that of the horizontal seams,
it should be strictly classed in a district to itself, but as the coals
are the same as those found in the nearly horizontal strata southwest of Dayton and in the Raccoon Mountain, I have thought best
to class all under one head, making merely a sub-head of horizontal
seams.

Again, the great development of Upper Measure coals on the
waters of Coal and Poplar creeks, are undoubtedly the creation of
one era of the carboniferous period, but the routes by which they reach
the markets are so dissimilar, that I have thought best to class them
under different heads. I have further adopted the plan of classing
as the "East Tennessee Coal-field" all the area east of the Sequatchie Valley on the southwest, and a line therewith northeast to the
Kentucky line, using as a distinction Chattanooga division and
Knoxville division. I have classed as the Plateau district all those
coals found north of the divide, between the waters flowing into
the Cumberland and those flowing into the Tennessee. This should
properly include the Sparta region, but as these coals seek a market
entirely different from the others farther east, on the Cincinnati
Southern and the Elk Fork, they will be classed in the "Middle
Tennessee Coal-field. The Elk Fork country contains some of the
upper coals, and is reached from Knoxville, hence will be classed
under that division.

Another mode of classifying these coals, which is also adopted in
other States, and in mineralogy is by characteristics known as
coking and non-coking. While some of our coals are better adapted
to making coke than others, yet none can be said to be so specially
adapted to that purpose as to be very poor for any other use.
Some are superior to others for use in the blacksmith shop, in fact,
may be said to have special value for that purpose. The hard
block coals of the sub-conglomerate measures are so much more

valuable for steam and grate use, and also stand transportation so well that those may be said to be their special use, at the same time our best coking coals are excellent for steam and grate purposes, combining a series of good qualities unsurpassed by any other coals in any section.

The classification I have adopted is as follows:

EAST TENNESSEE.—Knoxville Division.—Elk Fork district, Careyville district, Coal Creek district, Elk Fork district. Chattanooga Division—Poplar Creek district, Plateau district, Walden's Ridge district (inclined and horizontal).

MIDDLE TENNESSEE.—Sewanee district, Sparta district.

EAST TENNESSEE.

KNOXVILLE DIVISION.

The area which contains the coals of the upper series commences about eight miles northeast of the Little Emery river, and extends to a point about ten miles southwest of Cumberland Gap, the northwestern boundary being somewhat like a half circle. The southeastern is the peculiar outlying ridge I have mentioned as Walden's Ridge, which has an almost continuous northeast and southwest direction. This area comprises the Poplar Creek district, the Coal Creek district, and the Careyville district.

In the immediate outer vale of this area, the Walden's Ridge, the inclined coals are to be found, but they are not now worked at any point in East Tennessee southeast of the Emery river.

The Elk Fork district has not been thoroughly examined, and perhaps geologically belongs to the Plateau district of lower and middle coals, but in the division as to transportation, it is tributary southwards to Knoxville, and hence will be classed in that division of the East Tennessee coal-field.

ELK FORK DISTRICT.

This district comprises an area north and west of the Pine Mountain fault, and tributary to the Ohio division of the East Tennessee, Virginia and Georgia Railroad. Its southwestern boundary is the ridge dividing the waters of the Cumberland and Tennessee, and it is really an extension of the Plateau district, with more coals than are there found, the coals farther west being those of the lower series, which in Jellico are beneath the water's level.. The principal openings in this district are in Jellico Mountain, about sixty miles from Knoxville, five miles from the State line, and one hundred and eighty miles from Louisville.

This is, without doubt, to be one of the most important coal districts of the Tennessee coal-field. The Pine Mountain fault acting vigorously to the northeast, has there caused the preservation of some of the Upper Measures. In the measures as here presented are found seven seams of coal, only two of which are now known to be over three feet in thickness. These two are respectively thirty-six and fifty-two inches thick, the latter being the highest up the mountain. These seams all rise slightly to the northwest, and also have an upward strike to the southwest, the location of the workable coals commencing about ten miles from the point of the summit of the fault at Elk Gap. From thence these coals are continuous to the Ohio River through eastern Kentucky. The excellent facilities of transportation afforded by a well constructed trunk line of railroad, tapping the large cities of the West through two channels, and from its southern outlet radiating to three different sections of the South, with the proximity of the coal to that railroad affords unusual facilities for the transaction of a large business.

THE STANDARD COAL AND COKE COMPANY,

One of the operators, is composed of Tennessee capitalists; President and principal owner, E. E. McCroskey, Knoxville. The opening is three fourths of a mile from the main line of the railroad at Newcombe Station. They have 1,400 acres of land underlaid by a seam of coal four and one-half feet thick. The mine is well opened. Mr. McCroskey states that he has opened the lower seam, three feet thick, and that it makes a much better coke than the upper seam.

THE JELLICO COAL AND COKE COMPANY.

Is composed of capitalists chiefly from Lexington, Ky. Their opening is one and three-fourths miles from the main track, and they have entries driven so that as soon as transportation and demand permit, they can easily mine and ship 200 tons per day. They have about 1,500 acres underlaid with the four and one-half foot seam. Five other seams have been opened in this district, but of less importance than the seam worked as above.

The coal from the main seam, as analyzed by Dr. Peter, of Kentucky, gives:

    Moisture.. 2.36
    Volatile matter..36.44
    Fixed Carbon..60.60
    Ash (salmon color).. 1.60
    Sulphur... 1.16
    Fuel ratio...1 to 1.93

There is a slate parting in the middle of the seam where the mining is done. The dip is slightly upwards to the northwest.

The Jellico Coal & Coke Company report that in 1882 they paid out $5,000 for wages and $2,000 for mine supplies; that they have employed fifty hands, and paid $2.50, $2.75 and $3.00 for driving entries. The main entry is 700 feet long, by 7x6 in size. No coal shipped yet. The ordinary country schools and churches are in the vicinity. No liquor sold within three miles. The officers of this company are: Col. Sam L. Woolridge, President; Mort Mitchell, Secretary and Treasurer; Bret. R. Hitchcraft, General Agent; Jas. W. Fox, Superintendent, Lexington, Ky.; and Horace E. Fox, Assistant Superintendent, Newcombe, Campbell county, Tennessee.

THE EAST TENNESSEE COAL COMPANY,

which has been operating a mine at Coal Creek, have also opened mines in this district, immediately at the State line.

A number of other openings have been made on a small scale in this region, chiefly for prospecting.

CAREYVILLE DISTRICT.

Though at present of little note in the sum total of coal mined and shipped, yet this area, which is really a continuation of the Coal Creek section, will at no very distant day be of considerable importance. Only one mine is now worked there, and it to a small extent. The owners of this mine, known as the Campbell County Coal Company, deserve great credit for the pluck and perseverance they have shown in their operations.

The seam formerly worked was at the foot of the mountain and was very irregular, hence costly to operate. This seam was abandoned in 1881, and an entry commenced on a seam about 200 feet (vertical height) higher up the mountain. In this seam a good coal three and one-half to four feet thick has been reached, and gives every appearance of being persistent with a possibility of increase in thickness. From the fact that their work has been rather of exploration, but little coal has been shipped.

The Campbell County Coal Company lease from the Wheeler Iron & Coal Company, paying one cent per bushel royalty; shipped in 1882, 2,000 tons, valued at $14,000 at the mine; have employed twenty-one hands, and paid out in wages $9,500; selling price of coal at the mine has averaged seven cents per bushel; for mining coal, three cents per bushel was paid; for driving entries, $3.00 to $3.50 per yard; the average daily earnings of the miners was $1.50; the main entry is 800 feet long, 7x5½ feet; free school has been taught in the neighborhood for four months; the cost of living has been about thirty-three and one-third per cent above the average, and rents a little higher than formerly. H. P. Stone, Agent, Postoffice, Careyvile.

In the neighborhood of Careyville, but in the series of pitched seams, Messrs. Queener and Geers opened a seam of coal. It had been formerly worked, and a slope driven down on the seam. The gentlemen named extended the slope and operated the mine for a short time. The slope went down at an angle of 30° for about 210 feet, the seam then went off nearly horizontal for a short distance, and then pitched down again for twelve feet; it then became horizontal for about forty feet, when a slight dip was encountered, and at the bottom of it the mine was abandoned. The coal taken from it was of good quality, and it is to be regretted that the operators had not the means to go farther. The mine opening is in 200 yards of the railroad. This seam is undoubtedly the Sewanee, but having the pitch given it by the Walden's Ridge fault; the ridge itself, however, being here nearly gone.

Between the town of Careyville and Elk Gap, near the line of the railroad, a number of openings have been made on the land of the East Tennessee Coal & Iron Company. In two miles of Careyville one of these shows a 26-inch seam of excellent coal. This railroad follows the course of Cove Creek, which runs in a north and south fault, hence near it the strata are more or less disturbed, especially near Elk Gap, but at a distance of about half a mile to the west this disturbance ceases, and the mountains rise so as to contain better and more regular coal seams. For seven miles on the east the Fork Mountain fault exists, and the coal seams, if existing, have not been found. No regular work has ever been conducted on this company's land, but they have lately reorganized under the management of Mr. A. L. Maxwell, of Knoxville, a gentleman of great energy and address, who is rapidly settling up contesting claims, and this large and valuable property will soon be open for operations.

The peculiarity of the strata is very well illustrated by the following cut, taken from a report on this region made by Prof. J. P. Lesley. The point at which it was taken is about seven miles from Careyville, and there the strata have so subsided that the Kennedy seam, called Careyville coal, has sunk beneath water level. This is again brought up, however, as Elk Gap is approached, by the line of the Sequatchie Valley fault. At the point at which the section is made the railroad is built just on the right bank of Cove Creek. If valuable coals are found in this section the facilities for handling and shipment will be equal to those of any mining regions. The section is as follows:

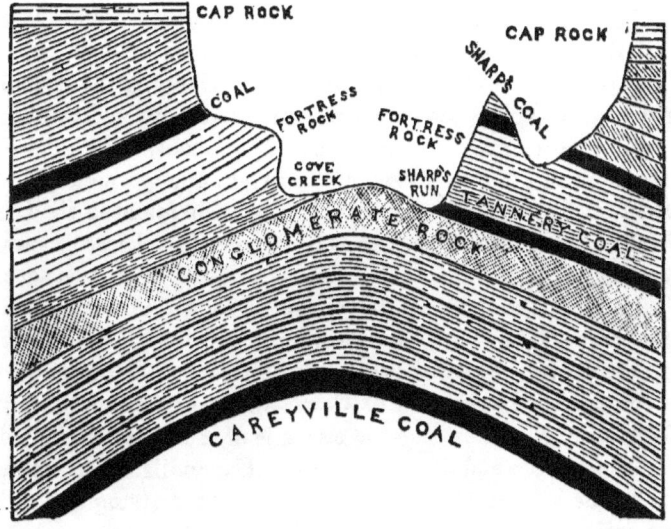

A mine was formerly worked in Careyville, on the Carey lands, but apandoned some time since, though by no means worked out. There are also other seams on this property, which is now owned by Mr. Rothschild, of Detroit. The seam formerly worked at Careyville is probably the same as the Sewanee. In 1871 and 1872 about 20,000 tons of coal per annum were shipped from Careyville. The old mines are in what is called the Kennedy seam, it had every characteristic of the Sewanee coal. The old Carey mine could yet be worked with success if properly opened, and a large amount of coal could yet be obtained from it. The seam noted heretofore as now worked by the Campbell County Coal Company, is about three hundred feet higher, vertical elevation. There is here, without doubt, a valuable series of coals which have passed unnoticed, as the locality has unjustly gotten into bad repute on account of the erratic manner in which the mining operations have been conducted.

Eight miles from Careyville a good seam of coal has been opened on G. W. Sharp's land, on Pine Mountain. This seam is six feet thick, with a parting of fire clay six or eight inches thick. The upper seam is three and a half feet thick, and is a hard cubical coal of excellent quality; then comes the fire clay, and below a seam of cannel coal. This is a valuable bed of coal, and the clay parting is an excellent material for a "mining," that is, a valueless matter to cut away so as to give room to prize the coal down. The cannel could be prized up, if found sufficiently valuable. The roof is good. Below this seam there are four other seams of coal. In the creek at this point and in the narrow valley adjoining the strata are nearly vertical, but in the ridges and mountains on each side they are nearly horizontal. Sharp's coal is probably continuous in the mountains on the west side of the creek, though no doubt at a different level; no coal, however, has yet been found in the Cumberland field corresponding with it in the peculiar characteristics of the seam.

The pitch of the strata in the creek is shown in the following section taken from the before-mentioned report of Prof. Lesley. Sharp's main coal is seen above the "Fortress" rock on the right. This creek is the largest fork of Cove Creek, it rises in a few miles of Careyville, flows north on the east side of Fork Mountain, joins the other branches and then runs south to Careyville. The uplifted strata on the left represents Fork Mountain here giving out. The whole region contains many wonderful freaks of nature.

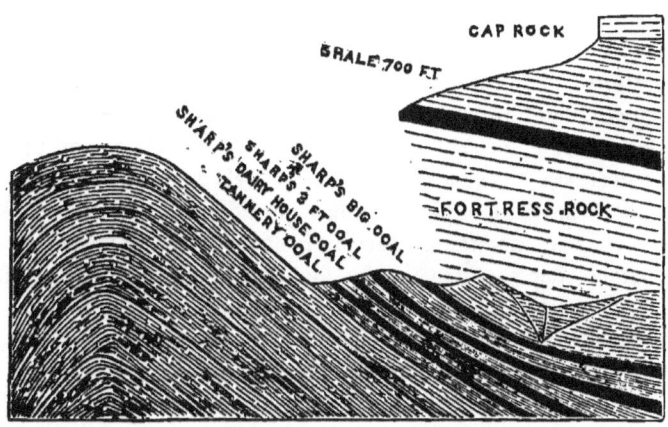

Prof. Lesley found on the west side of Cove Creek a seven foot seam, and also two others respectively three and four feet thick. A practical miner who has studied this country, considers this thickness of seven feet as merely local, but that a good five foot seam may be found. Prof. Lesley thinks these seams belong to the Upper Measures, and that when opened, may be relied upon to furnish a permanent and regular supply of coal. The downthrow which commences near Careyville, reaches in the distance to Elk Gap, a depression probably as great as that of the more southern measures at Cove Creek and Winter's Gap, and hence the upper coals may be here preserved as at those points. It is possible that the Sharp coal may be below the surface on the west side of Cove Creek, or that it may be identical with the seam found by Prof. Lesley, and thought to be seven feet thick. The disputed position of these coals make it very difficult to identify them with any others in the field farther to the south.

THE COAL CREEK DISTRICT.

The stream from which this district derives its name, unlike its neighbor on the southeast, comes through Walden's Ridge by a single channel, and inside the gap its forks are merely one to the right and one to the left, hence this region has not such advantages for a great output of coal as the Poplar Creek region, which will be treated of hereafter. . The main mountain also abuts up to Walden's Ridge, being only separated from it by the narrow valley of the creek bed; hence, from its proximity to the disturbed strata, or rather to the line of the dislocation, its influences have had some effect on the horizontal measures. The first or lateral area on the two forks of Coal Creek, available for coal openings, is not more than ten miles, and to the northeast the strata are much disturbed, while on the southwest but little or no exploration has been done. The streams running westward into the mountain are short and their sides precipitous, hence offer poor accommodation for mining plants; therefore all large operations must be made on the front. The great drawback to this district, and which will work to its injury when other fields are opened to transportation, is the excessive royalty charged for the coal in the ground. This is twenty-five cents per ton, or one cent per bushel, being more than twice as much as the average royalty of the whole United States, that average being fourteen cents per ton. The following is the average royalty in the various States, derived from the census returns:

Alabama	16
Georgia	12½
Illinois	9
Indiana	15
Iowa	23
Kentucky	14
Maryland	11
Ohio	17
Pennsylvania	14
Tennessee	25
Virginia	13
West Virginia	15

The reason why this excessive royalty is maintained: The Coal Creek Mining & Manufacturing Company owns nearly all the coal lands on Coal Creek. The stock of this corporation is owned, one-half by a gentleman in New York City and the residue by various parties. The royalty now yields about $40,000 a year income, of which, taxes off, the New York gentleman gets full $18,000, and yet has never spent a dollar to improve or develop the property. Of course he does not intend to lessen his income until he is forced to do so, and the other stockholders cannot make any change of policy.

This excessive royalty will in the end result to the benefit of the State, as it will cause the opening of other mines, which could have been prevented by a more far-sighted and wise policy. The mines of the Careyville and Elk Fork districts will be effective rivals, especially as the railroad has proposed to take coal from them to Knoxville at very near the same rates as from Coal Creek.

This excessive royalty also works to the injury of the land company itself, as the operator brings out only such coal as will sell to the best advantage, and works the mines on the cheapest plan he can with safety to his men and property. The land-owning company constantly speaks of the railroad freight rates as too high, but the railroad has repeatedly stated that when the royalty is reduced they will reduce rates. As will be seen further on in my report, as far as the Southern market is concerned, this stubbornness is likely to create a successful rival on Poplar Creek.

There never has been any accurate and full survey of the Coal Creek district, nor in fact any part of the East Tennessee coal field. The only data from which information can be obtained, is from a brief report of a reconnoisance made by Prof. Bradley, which from that gentleman's well-known ability as a geologist as well as care in conducting this work, may be assumed to be at least approximately correct. From the bottom of the creek at the railroad bridge, just at the foot of the creek, and inside the Walden's ridge, where the horizontal strata commence, to the top of the Big Butte Peak, Prof. Bradley found eighteen seams of coal. All these in the horizontal strata above water level, while in the pitched seams of the Walden's ridge, he found three more. Of these eighteen seams he thinks there are eight two and a-half feet and over in thickness. The seam now worked, he classes as seam E, and is the second above the creek. It is probably the equivalent of the noted Pittsburg and Youghiogheny seam. This seam may be always relied upon for four feet of good coal, though the area of coal and shales frequently reaches six feet. The shales and thin coals are in the top, there being usually about four feet of coal on the bottom in two bands, separated by one to two inches of shale. The mining is usually done on top and prize up, especially in the Knoxville Iron Company's mine, some of the others mine underneath and throw down. Powder is but little used except in entries.

As the use of this coal has been chiefly for domestic purposes, only two or three of the mines are worked the entire year, and those are operated with a reduced force over the winter months. For this reason the price of mining is greater than if the miner could make every day in the year; he must make enough in one hundred and fifty or two hundred days to take care of such part of the other one hundred that he may be out of a job. The use of this coal for steam generating, both in stationary boilers and in locomotives, is rapidly increasing. The East Tennessee, Virginia and Georgia railroad first made experiments for using it in their locomotives in *1875; now it is used by them entirely, is used largely by the Norfolk and Western, the Richmond and Danville, the Georgia Central, and the Georgia railroad, and negotiations are in progress for its use on South Carolina roads. But the great success of a coal mining operation is in supplying iron furnaces and manufactories. The only iron works at present supplied solely from Coal creek is that of the Knoxville Iron Company at Knoxville. The foundries at that city get most of their coke from Etna, and not a single iron furnace is tributary to the Coal Creek district for coal or coke.

The excellent quality of the Coal Creek coal has been attested by many years of use, and though the roof over the coal is good, and mining can be conducted very cheaply, yet it is probable that the limit of production there has been reached and is more likely to fall back than to increase. Its development before other sections was caused by the construction of the Knoxville and Ohio to that point and its failure for many years to be continued more than a few miles beyond. This road has been and is always likely to be its only outlet, and though it is popular to complain against its managers, yet they are as liberal in freight rates as the owners of the coal lands in royalty. They charge less for the use of a property for which they paid hundreds of thousands of dollars, on which they pay taxes to State and county, and for the improvement of which they annually expend thousands of dollars, than the Coal Creek Mining and Manufacturing Company does for the use of their property which cost them but little more than the conduct of some lawsuits, on which they pay comparatively low taxes, and expend nothing for improvements, and which property to-day would be worth but little, had the Knoxville and Ohio road gone through Big Creek Gap, where it was intended by its original charter to have gone. The connections of this road under the same management into Georgia, North Carolina and Virginia offer great facilities for reaching various markets. The reports of the operators sent in to me state that freight rates are lower than two years ago.

Under a law enacted by the Legislature, forbidding the sale of intoxicating liquor within four miles of an incorporated school, such sale is kept away from the neighborhood of the mines. There are two churches in the neighborhood of the mines; a free school is taught four months of the year, and a subscription school for ten months.

The Coal Creek mines were first opened for shipping coal on the completion to that place of the Knoxville and Ohio railroad in 1870. The annual shipments have been as follows:

YEAR.	TONS.
1871	36,000
1873	46,006
1874	36,816
1875	62,369
1876	57,459
1880	150,000
1882	200,000

An increase in ten years of over 500 per cent, and estimating the value at the mines at the average of seven cents per bushel or $1.75 cents per ton has increased from a valuation of $63,000 in 1871 to $350,000 in 1882. In 1871 only three mines were actively at work on Coal creek, part of the coal for that year coming from Careyville, and they employed about one hundred hands; two more were preparing to operate. Now there are seven employing five hundred hands, and giving support to fully five times as many persons. At the same time to feed the operatives and the stock used in the mines, thousands of dollars worth of provisions and provender are annually bought from the farmers. For a short time there was a depression on account of the employing of convicts, and the general apathy of business, but now the town has recovered, many new houses have been built, both for dwellings and business, and the general appearance of the place indicates that its people are prospering.

The mode of mining and interior arrangement of all the mines on Coal Creek is nearly the same. The system of ventilation is by an entry parallel to the main entry, with fire and flue near the entrance. Only one mine has a double sized main entry and a double track therein, all others are seven feet wide by 5 to 6½ high. The cross entries are cut every thirty-six yards, usually seven feet wide, five to six feet high. Pillars thirty-six feet wide are left between the entries and between rooms, the rooms twelve to fourteen yards wide by twenty-four yards long.

The following is a detailed statement of the mines:

THE KNOXVILLE IRON COMPANY

During the year 1882, shipped 98,645 tons of coal to various markets in Southwest Virginia, North and South Carolina, Georgia and Alabama. The value of the coal shipped was at the mines $135,437.57. The price of coal ranged from five to eight cents per bushel. The amount invested in mining plant is $50,000. The company employs as high as one hundred and sixty men, of whom one hundred and thirty are convicts; the average number employed during the year has been one hundred and twenty-five. The sum of $58,854 was paid out in wages, and $10,164 for mine supplies. Their main entry is fourteen feet wide and three-fourths of a mile long, the side entries are 7 x 5 feet. Free labor is paid three cents per bushel for mining, and the cost of mining has increased about twenty-five per cent over 1880. Provisions and rents are also higher in price. The average earnings of the miners is $3 per day. The company works their convicts and some free labor all the year, but works a full force, full time only one hundred and fifty days. W. R. Tuttle, of Knoxville, is president, Capt. Jno. F. Chumbley in general charge of the mines, Jno. L. Davis, assistant, and John Hightower has charge of the mining operations.

This company conducts the largest operation on the creek, and its business is very well managed. Its facilities, both inside and out for handling coal, have capacity for producing a larger quantity, at less expense than any other company at present on Coal creek. Their entry, however, is somewhat unfortunately located, but probably as good as could be obtained from the area of land which they were allowed to work. Their main entry is 1,300 yards long, and is of double size, so as to admit two tracks the whole distance. The ventilation is accomplished by a parallel entry, and is very perfect. This parallel entry is also arranged to serve as a second outlet to the mine, in case of accident. In this entry a deep sump has also been cut to which all the water of the mine will flow. The seam dips for some distance on the main entry into the mountain, at the rate of two inches to every eight yards; hence the necessity of the sump alluded to, and also of a steam engine for pumping. This water was formerly a great inconvenience, but the present drainage will be effectual and permanent, thereby taking off one item of expense. This mine is well opened and the underground arrangement well planned. About fifteen per cent of the amount cut down by the miners is slack, which is not brought out but is raked back. There is therefore in this mine not less than 100,000 tons of waste slack, which but for the excessive royalty might be put to some use.

This company's mines are worked somewhat differently from the others, their main entry is fourteen feet wide and six feet high; air way seven feet wide and five feet high to higher; cross entries are seven feet wide by five feet high; the rooms are worked three hundred feet long by forty-one feet wide. The pillars on the side of the main entry are thirty-six feet thick, between the rooms the pillars are left twenty-one feet thick. The mining operations of this company are very well conducted, and its executive management of the best character.

This company had January 1st, 1883, 130 convicts, though their usual number is only 100; their number of free laborers range from thirty to fifty. They work their mine all summer though with reduced force. The convicts are leased from the lessees of the Penitentiary, and the price paid is private. It is estimated, however, that the total of leasing, feeding, clothing and guarding amounts to $1.20 per day for each convict. Their task in digging coal is 100 bushels per day, and most of them easily get through, even in winter, before dark. They make some money for themselves by extra work, and also by making trinkets out of the thin cannel coal, which they sell to visitors. Their quarters are clean and comfortable, and they are evidently well cared for, it being clearly the pecuniary interest of the company to have them healthy and able to do work. They are supplied with abundance of fresh meat in winter, and vegetables in summer. Dr. Smith, a young physician of great skill, is employed by the company, is always in attendance, and it is a part of his duty to inspect their quarters every day. Capt. Chumbley says his idea of the best way to cure disease is to prevent it, and if it comes, to attack it in its earliest stages.

THE BLACK DIAMOND MINING COMPANY

Have large tracts of land leased, some of which is again subleased, but they ship coal from only one mine, known as the Black Diamond. In 1882, they shipped 30,000 tons of coal to various points in Tennessee, Georgia, South Carolina and Mississippi. The value of the coal shipped was $40,000 at the mines. They worked fifty hands whose usual earnings were $2.25 cents per day, and paid out $25,000 in wages and $10,000 for mine supplies. The price of coal at the mine ranged from five to eight cents per bushel. The price paid per bushel for mining is three cents; for driving entries $2.75 to $3.25. Their main entry is one-half a mile long. The usual full time worked is 200 days. Freight rates are lower than two years ago, and the cost of mining has increased. Capt. B. F. Rooney is superintendent. The mine is very well ventilated, has two outlets and has natural drainage. The shute and outside track arrangements are excellent, the shute and tipple especially. T. H. Heald is agent, Knoxville Tenn.

THE STAR COAL COMPANY

In 1882, shipped 24,000 tons of coal to Chattanooga and Knoxville, Tennessee, Atlanta, Rome and Augusta, Georgia, Greensboro and Salisbury, North Carolina, which was valued at the mines at $42,000. The average selling price at the mine was seven cents per bushel. The number of hands worked was eighty, who earned (miners) from $2. to $3. per day, and outside men $1. to $2. The total amount of wages paid was $30,000. From two and a half to three cents per bushel is paid for mining, and $3. to $5. per yard for driving entries. The main entry is 9 x 6 feet and 275 yards long; side entries are 7 x 5½ feet. They have an entry parallel with main entry for ventilation and to make a double outlet to the mine. This mine has only been in operation one year. James Frayser is superintendent, and Thos. L. Moses, Knoxville, is president.

THE ANDERSON COUNTY COAL COMPANY

In 1882, shipped 11,000 tons of coal valued at the mines at $17,000. The markets were East Tennessee, Virginia and Georgia. The average price of coal at the mines was 5½ cents per bushel. The employees numbered twenty-five, who earned an average of $2.50 per day, and the total wages paid during the year were $11,146, and $551 was paid out for mine supplies. For mining three cents per bushel is paid, and for driving entries $3.25. Their main entry is 7½x5 feet, and is 600 yards long, side entries the same size. The company has a store, and sells about $6,000 worth of goods per annum. Chas. McKarsie is superintendent.

THE CENTRAL COAL COMPANY

During the year 1882, shipped 23,550 tons of coal to various points in Tennessee and Georgia, the value of which at the mines was $35,375. The number of employees was seventy-five, the miners usually earned $2.50 per day, and $15,662 was paid out for wages. The company keeps a store, but do not report any purchases for that purpose, or for the mine. The selling price of coal at the mine is reported at $1.35 to $1.75 per ton. The price paid for mining was two and a half cents for mixed, and three cents for lump coal; from $2.50 to $3.75 per yard was paid for driving entries. The entries are 7x5 feet, and they have 1,048 yards of entry. Full time was made only one-half the year. This company leases the old Franklin mine from the Black Diamond Coal Company, and are drawing the pillars in their old mine, preparing to abandon it. This old mine had a coal area of about fourteen acres, and they obtained from it about 70,000 tons of coal. W. B. H. Wiley is superintendent, Coal Creek, and Col. E. E. McCroskey agent, Knoxville.

THE EAST TENNESSEE COAL COMPANY

In 1882, shipped 11,760 tons of coal to Atlanta, Macon, Knoxville, Bristol, and other points in Tennessee and Georgia. The value of this coal at the mines was $18,000; the selling price averaged from five to eight cents per bushel; the number of employees was forty, and the amount paid out in wages $10,820; the amount paid out for mine supplies being $2,000. The company does not keep a store. The price paid for mining coal was, for mixed two and a half cents, lump three cents, and from $1.25 to $2.50 per yard for driving entries. The main entry is 700 yards long, 7 x 5 feet. During 1882, about 175 days of full time have been worked. This mine has quite doubled the number of employees and their product in the last two years. Considerable improvements have also been made in the mine. E. J. Davis is the superintendent, and Capt. Jno. M. Brooks, president and agent, Knoxville.

THE COAL CREEK MINING COMPANY.

Has been actively engaged during the year driving a main entry into the heart of the mountain, so as to reach the area of a large lease it controls. This entry was driven through the old workings of the Empire mine, which once held an area on the face of the mountain, and therefore was not only costly but dangerous, as the pillars had long ago been drawn, and the space once occupied by coal was a mass of "gob" of the worst character, a mixture of old props, slack, slate and fallen roof. A first-class entry has been driven through this, lined on both sides with cribbing or "lagging," made of the best of white oak timbers about the size and length of railroad cross-ties filled in with stone, and where necessary the roof has been supported in equally as firm and substantial a manner as the side timbering. This entry is the best piece of work on Coal creek, and probably in the south. In fact I cannot see how it would be possible to surpass it in any section. It is a perfect straight line running N. 75° W., which is at right angle to the face of the mountain and of the creek, also of the general course of Walden's ridge. It is 750 feet long, 10 feet wide by 7 feet high ; therefore will readily admit of the use of a mine locomotive. At the time of my visit, it had been driven a distance of 725 feet, and the miners were cutting through the fault which had stopped the operations of the company which worked the old mine. This fault did not have the course of the mountain, but ran rather diagonally across the entry. The course as well as I could determine, being N. 75° E., and S. 75° W. And I may here add that while the general trend of the East Tennessee faults, and the valleys caused by them is N. 25° E., and S. 25° W., that there are also a series of cross faults or uplifts running across the strike of the other in a line N. 75° E., and S. 75° W., and frequently cutting off or rather crowding out the former. T. H. Heald, Knoxville, is President of this company.

HECK MINES,

During the year 1882, shipped coal to various points in Virginia, North Carolina and Georgia, as well as Tennessee, (amount not stated). The number of men worked was eighty, who earned from $1.50 to $3.00 per day, according to skill and industry, and the total amount paid out as wages was $27,868.51; there was paid out for store supplies $8,368.20, for miner's supplies about $8,000. The selling price of coal at the mine was from 5 to 7 cents per bushel. From 2 to 3 cents per bushel was paid for mining coal, and from $2.00 to $3.75 per yard for driving entries. The main entry is 670 yards long, 6 feet high, 8 feet wide; side entries 5 feet high and 6 feet wide. One man had his leg broken by the falling of slate, resulting from carelessness. This mine was opened and the first coal shipped in December, 1881. Eleven months of full time was made last year.

Mr. P. Thomas, Coal Creek, is Superintendent.

MARKET AND FREIGHT RATES.

The following are the freight rates on the ton of 2,000 pounds and distances to various points:

To Knoxville, 30 miles, $1.00 for domestic coal, 70 cents for manufacturing; to Chattanooga, 140 miles, $1.25 for all purposes; to Bristol, 116 miles, $1.80 for all uses; to Asheville, N. C., 144 miles $2.65; (30 miles of this over W. N. C. R. R., hence the higher rate, though less distance than to Bristol;) Goldsboro, N. C., 464 miles, $3.80; Raleigh, 414 miles, $3.60; Rome, Ga., 180 miles, $1.90; Selma, Ala., 378 miles, $3.00; Atlanta, Ga., 240 miles, $2.30 for domestic, $2.00 for steam and manufacturing coal; Macon Ga., 330 miles, $2.75; Milledgeville and Augusta, $4.00; Brunswick, $3.25, and Savannah, $3.50. Placing the coal at 6 cents per bushel, $1.50 per ton at the mine, and the cost of a ton in Brunswick would be $4.75; at Savannah, $5.00. The Westmoreland and Youghogheny coals, which are the same as Coal creek, on January 24th, were priced on barge in New York City at $4.75 per ton.

CHATTANOOGA DIVISION.

POPLAR CREEK DISTRICT.

Three of the streams which are the headwaters of Poplar Creek come through Walden's Ridge at three different points, thus making thorough cuts, inside the ridge these three streams are divided into numerous branches, which give easy routes far up into the mountains. In this respect it has the advantage of Coal Creek, where there is only one pass through the ridge. The portals through which these forks of Poplar Creek come into the valley are known as Donnavan's Gap and Winter's Gap. The latter is indiscriminately applied to two breaks or gaps in the mountain which are only about two hundred yards apart, but through which two distinct streams flow, and the region on the waters of which is separated by a high ridge running back into the main mountain. The former is two miles northeast of Winter's Gap, and the stream flowing through it is known as Mountain Fork. The streams at Winter's Gap are known as Salt Fork and Indian Fork. Seams of coal have been opened on all these creeks, but mining operations are now only conducted on Indian Creek.

The coal district comprising the area to be reached by the various forks of Poplar Creek has more available coal than any other district of the Tennessee coal field, and it is claimed that the coal is of better quality. There are known to be nine beds in the Brushy mountain alone, averaging over three feet thick, and as most of them have only been measured at their outcrops, there is no doubt that when worked they will be found to be thicker. And besides those there are the pitched seams in Walden's ridge, which have been opened and tested, and also a number below the level of the foot of the mountain which have been ascertained to exist from borings. The area of coal lands tributary to these gaps is but little, if any less than 500 square miles, and the facilities for reaching the coal seams from different directions afforded by the various valleys forming the beds of the numerous streams has not its equal in any coal area in the southern States, if it has in the United

States. Through these outlets a supply of coal could be brought to a main line which would tax its carrying capacity to the utmost. At present the only means of transportation is the Walden's Ridge Railroad, which to Oakdale Furnace and to Emery River, is a very well constructed narrow gauge, but now connects to the Cincinnati Southern by a branch which neither in construction or grades will permit a large burden of traffic. From Winter's gap to Oakdale this road is generally taxed to its full capacity to carry the coal needed at the furnace, and thence it can do but little more than carry off the iron made, therefore the prospects of great development of this district are not very encouraging. With a wide gauge railroad the product of coal from there would be very large, as the supply is inexhaustible, and the distance to market less than from Coal Creek.

Inside the gaps mentioned, are first comparatively low ridges; these extend back into very high mountains in which numerous seams of coal are piled on top each other in regular order, and with a general rise to the northwest, pass through the mountains to the side on the waters of New and Emory rivers, thus giving an immense area of coal which can be drawn upon for hundreds of years. These great mountains, a chain extending in fact over 40 miles in length, and rising at intervals in peaks from 2,500 to 2,800 feet above the level of the waters of Poplar creek, may be attacked by the miner at not less than seven different points which afford ample room for large mining plant, while their base on the main seam is not less than four miles to the outcrop of the seam on the waters of New river. It is evident that the area thus available is very great; at the same time in the ridges alluded to, there is another vast quantity of coal not taken into this calculation.

At the time of the survey of the Tennessee & Pacific Railroad, Dr. Safford made a report on the mineral wealth of this region, from which the following section of the strata in one of the mountains is taken:

Sandstone at top of mountain	100	feet.
Shales and sandstones	249	"
Coal	6	"
Shales and sandstones	240	"
Limestone	37	"
Shale	74	"

4

Coal	4	feet.
Shales	40	"
Sandstone	60 to 80	"
Shales	50	"
Coal [outcrop]		
Shales and sandstones	120 to 170	"
Shales	130	"
Shales with nodular iron ore	120	"
Coal [outcrop]	1	"
Shales	6	"
Coal [outcrop bed 5]	3	"
Shales and sandstones	110	"
Shales mostly	100	"
Sandstone	70 to 100	"
Shale	45	"
Coal [with four inch shale parting]	4	"
Shales and sandstones	180	"
Sandstone	50 to 80	"
Coal	3	"
Shale with nodular iron ore	25	"
Sandstone and shale	150	"
Coal [outcrop]	1	"
Fire clay	4	"
Shale	5	"
Coal	5 to 7	"
Fire clay and slate	5	"
Shale	30	"

[Foot of mountain nearly.]

"This section is a natural one, there having been no digging except in the lower bed of coal, all the other exposures being mere superficial outcrop. Notwithstanding, we have nine beds presenting an aggregate thickness of twenty-six feet of solid coal. If the beds were properly opened they would doubtless prove to be thicker. The aggregate must be at least thirty-six feet. Moreover, search with picks and shovels would reveal other beds now concealed. The coal beds below the level of the road at Winter's Gap are not represented in the above section. The lowest in the section is a splendid bed. The beds above are likewise accessible, and contain an amount of fuel of which we have no adequate conception."

The seam now worked at Poplar creek is the same as that known as seam E at Coal creek, and the only one there worked. This seam can be depended on for four and a half feet of good coal. There have been local openings of seven feet, but it is my opinion that in the main mountain it will be persistently four and a half feet of coal with slight slate partings, which will give the miner full five feet of working space. The lands in this district are in the ownership of various parties, not in the narrow compass of a few as at Coal Creek, and hence to it may be looked for a reduction of the enormous royalty which has been established at Coal Creek, and which will ever drive off the investment of large amounts of capital and inauguration of large enterprises. From large operations doing a great business on a small margin of profit can only be expected cheap coal, and cheap coal and cheap coke are now the needs of our iron manufacturing enterprises.

The value of this Poplar creek coal for coke-making has been fully proven by the Oakdale Iron Company, but will be more fully discussed hereafter. As a steam and domestic fuel, it has no superior among bituminous coals. The average quantity of ash from several analyses is a small amount over one per cent less than any coal I have been able to find in a long series of analyses of Kentucky, Ohio and Pennsylvania coals in my possession.

I have stated the route of transit for this coal as the Walden's Ridge Railroad. This road is chartered to connect with the Cincinnati Southern at Emery Gap, and to be built to the Ohio branch of the East Tennessee Virginia and Georgia. A route has been surveyed to Clinton on that road. The distance from the mines to Emery Gap is eighteen miles; from that point to Chattanooga is eighty miles, and the grades are so low that large train loads can be carried at a low cost of transportation. That is the route by which this coal will seek a market, and hence its classification has been placed in the Chattanooga division.

There are now six mining operations being conducted on the waters of Poplar Creek, all, however, in their infancy, and the work so far being more of preparation than for actual business. The Oakdale Iron Company mines extensively for its own use, and one other company does a small shipping business, the cause of this is stated above in the fact of limited transportation facilities. The names of the companies with some data of their operations is as follows:

THE OLIVER COAL COAPANY—
Have expended about $5,000 in opening mine, building shutes and track.

THE SOUTHERN COAL COMPANY—
Have expended about $4,000.

THE EUREKA COMPANY—
Have expended about $5,000.

THE MT. CARBON COAL COMPANY—
Have shipped from one to two cars per day during the past winter, but do not furnish data of number of hands or expenditure.

THE OAKDALE COMPANY—
Also did not furnish full information, hence only an estimate of their product could be made. Their operations were irregular; sometimes mining 200 tons per day, and again running up to 400 tons.

With proper transportation this is undoubtedly destined to be one of the most prosperous sites of coal mining in the Southern States. Not only is the coal excellent for steam, for gas and for domestic use, but the fact that it makes a first-class coke has been fully demonstrated. The fact, too, that the lands have a general ownership, and are not in the hands of a single corporation, will tend to create competition and a larger business.

The pioneer in the coal business of East Tennessee, and probably of the State, was Henry H. Wiley, of Anderson county, a native of Virginia, and a land surveyor by profession. He endured hardships and privations with perseverance and pluck, but died just as the bright visions, of which he had long dreamed, began to be realities. Unlike most pioneers, however, his children are now reaping a rich reward for his long years of labor. Mr. Wiley opened a mine on Poplar Creek, and for many years boated coal down to Huntsville and Decatur, Ala., on the uncertain tides of the winter months. He hauled the coal four miles to a point below the junction of the four forks of Poplar Creek, where it was put in boats, floated out that stream to the Clinch, then into the Tennessee and down the then perilous navigation to its destination. The day may yet come when this pioneer mode of transportation with improved facilities will again be the great route of coal carriage, and this coal find a market in St. Louis and New Orleans. It is certain that when the Tennessee, the Clinch, Emrey and Poplar rivers are improved, as are the Ohio, the Monongahala and the Youghiogheny, that the possibilities of reaching a Southern market from Tennessee will be more probable at all seasons of the year than now from the rivers above named in Pennsylvania.

The surveys made by the United States Government demonstrate that such a feature is by no means an impossibility. Col. Gaw, in his report states, that with proper improvement of the Muscle Shoals, coal should be delivered from Emery River to Paducah for twenty-seven cents per ton, and Maj. King states that the distance by his survey is 565 miles from Kingston to Paducah; it is 919 miles from Pittsburg to Paducah, and the coal must come to Pittsburg, down the Monongahela, by navigation, which is only reliable in high water.

THE PLATEAU DISTRICT.

The area which I class as the Plateau District, has been stated; it commences on the Cincinnati Southern Railway, just south of Sunbright Station, the line of boundary there being the dividing ridge between the waters of White Oak, which flows north into the Cumberland, and those of the Emory River, which flow into the Tennessee. This ridge is a great anticlinal uplift, the same which has broken and formed the Sequatchie Valley on the south and the Elk Fork Valley on the north, but not in either of these sections forming the water divide, as it does on the line of the Cincinnati Southern Railway, and immediately at the Elk Gap Tunnel on the Knoxville & Ohio. Through the counties of Van Buren and Cumberland the strata of the water-divide ridge are apparently horizontal and underlaid with good seams of coal; these are evidently gone at Triplett's Gap and Tunnel No. 10 on the Cincinnati Southern Railway and probably nothing but the thin seams of the Lower Measures are to be found. At the same time it must be admitted that the geology of this region has not been by any means thoroughly studied. But the water-divide on the east of the

Cincinnati Southern line deflects to the east and again contains seams of good coal even into the Upper Measures, through the counties of Morgan and Anderson into Campbell. In the latter county, at Elk Gap, the measures are again greatly disturbed, the synclinal fault there striking the water-divide, but passing thence to the north-northeast, while the water-divide runs more nearly east, and contains good seams of coal, until it is lost by a union with the Walden's Ridge fault at Cumberland Gap. The line of the great Sequatchie Valley fault is seen, therefore, to be regular in its direction and its greatest action to have been on the south-southwest of one extreme and the north-northeast of the other, that its action and the Walden's Ridge downthrow seem to have been in concert, and to the former is due the Cumberland Plateau as the latter is the undoubted cause of the preservation of the great area of Upper Measures in Morgan, Anderson and Campbell counties. Also to the drop of the strata on the northern side of the strata of the anticlinal is to be attributed the preservation of the Sewanee seam almost constantly along the mountain on the northern side of the Sequatchie Valley, also the steady decrease of the thickness of the strata above that seam to the north, and its entire absence before reaching the Kentucky line.

Near Triplett's Gap, on the Cincinnati Southern Railway, the country is somewhat rolling, and so continues to beyond New River, but nowhere with the high mountains which characterize the region to the east. In this section two mines have been opened and extensively operated during the past two years.

THE CROOKE COAL AND COKE COMPANY

Operate near Glenmary Station. The mine is about seventy-five feet higher than the railroad track, and seven thousand and five hundred feet distant. This distance is traveled by a narrow gauge track three thousand and five hundred feet long, with a slight down grade, on which one mule pulls easily twelve to fourteen mine cars, this track reaches to the tipple, which is a very complete piece of work, probably the best constructed and most perfectly arranged in the State. The coal is sorted into large lump for domestic use, small lump for steam, and slack, the latter is made into coke. The seam of coal worked may be said to be constantly thirty-four inches thick, of this there is a peculiar shelly coal on top, ranging from four to eight inches thick, this coal makes their slack ; otherwise the coal of their seam is very compact and bears transportation with very little breakage. Their mine is very well ventilated and has a number of outlets. The elevation of this coal seam is 1,462 feet above the sea. Its geological position has not been positively determined. Mr. Jno. H. Clark, the Superintendent of the mining operations, is an excellent Manager and good Engineer.

For the year ending December 31, 1881, 45,000 tons were shipped to various points from Lexington, Ky., on the north, to Augusta and Macon, Ga., in the South. The average number of hands worked during the year was 175, the earnings of miners ranged from $2.00 to $4.50 per day ; the price per bushel paid for mining being three and a half to four cents. The selling price of the coal at the mine ranged from two cents per bushel for slack to eight cents for large lump. The main entry is three thousand and five hundred feet long, and there are seven thousand and five hundred feet of side entries, all seven by five feet. The mine was opened in September, 1881, the usual full time worked is five and a half days per week. There are the usual common schools in the neighborhood, no churches near by, and liquor is sold abundantly around the railroad station. Geo. W. Darnall, Lexington, Ky., is President, Jno. H. Clark, Glenmary, Superintendent.

THE HELENWOOD MINES

Are at Helenwood Station, on the Cincinnati Southern Railway. In the fall of 1882 the company formerly operating, failed, and after being in the hands of a receiver for some weeks they were leased by A. B. Stone & Co., of Chattanooga. These gentlemen state that they have been operating so short a time that their report would be so meagre as not worthy to publish, and that the books of the former company are not in their hands. A. C. Eaton, of Helenwood, is Manager, and F. I. Stone, of Chattanooga, Agent.

An analysis of these coals has been furnished as follows :

	Glenmary.	Helenwood.
Water	1.67	1.83
Volatile matters	34.53	41.29
Fixed carbon	61.66	54.24
Ash	2.15	2.64
Sulphur	0.50	......
Fuel ratio	1:1.78	1:1.31

Both these mines ship north and south over the Cincinnati Southern Railway. The freight rates are liberal.

From the Crooke Coal Company's mines to Chattanooga is 112 miles, to Lexington 144 miles, to Cincinnati 223 miles. From Helenwood to Cincinnati 200 miles, to Chattanooga 135 miles. From Oakdale Junction to Chattanooga 82 miles.

WALDEN'S RIDGE DISTRICT.

INCLINED STRATA.

The peculiar position of the strata in what is known as the Walden's Ridge, from Cumberland Gap to a few miles south of Dayton, has been previously alluded to, and, as stated, this peculiar freak of nature in placing the strata exists over a long area of country. Probably the most accurate estimate of this region was that placed upon it by Prof. Lesley, formerly Assistant Geologist of Kentucky, and now Geologist of Pennsylvania, in the excellent and expensive survey of that State just now being completed. He determined that there had been in years gone by a great drop of the strata, a downthrow it may be called, whereby the interior basin of coals had been preserved from the erosions which had swept them away in other sections, and the Lower Measures had been cast many feet under the water level. Prof. Lesley's connection with the geological survey of Kentucky had given him some idea of this peculiar region before he examined it for the East Tennessee Coal and Iron Company, but such knowledge had not extended to the Dayton region, hence probably he did not know that there the great downthrow ceased to exert its influence.

The main seam of this Walden's Ridge upturned strata is undoubtedly the same as the Sewanee and Soddy, but from the peculiar compression of the strata, the other seams do not show at Coal Creek as at Etna, Daisy, Soddy and even at the Emery mines. There can now be but little doubt on the part of any one that these pitched strata go down under the horizontal strata of the Cumberland Mountain at Winter's Gap and Coal Creek; and the probability is that they rise with the Sequatchie Valley fault. The difference of elevation between the valley level at Rockwood and the appearance of the same seam of coal on the west side of Crab Orchard, (it has unfortunately not been searched for on the east side) is not more than would make a very gradual rise in a coal seam. Fifty feet to the mile is a very small rise, hardly perceptible, yet in ten miles it is a rise of five hundred feet, while the difference of elevation between what might be called the valley level of the coal seam at Rockwood, and what it should be at Crab Orchard is about 800 feet. Hence in the distance between the two places it is easy for the seam to rise to a corresponding elevation at Crab Orchard The strata of this ridge belong to the umbral or sub-conglomerate, and to what I have designated as the Middle Measures, the Lower of Pennsylvania. On the eastern slope of the ridge, some of the mountain limestone series frequently come up a short distance, this becomes more especially the situation as the Virginia line is approached; at Cumberland Gap, for instance, the limestone is far up on the eastern side of the mountain, and even the Devonian black shale is nearly one-half the way up. At the eastern base of this ridge a series of knobs is found all the way from Cumberland Gap to Chattanooga, in which the strata are pitched at the same sharp angle; these knobs are of the Clinton (New York) or Surgent (Pennsylvania) formation, and carry throughout the length named one or more seams of red fossil iron ore. Thus bringing one of the most valuable ores of iron in close conjunction—a few hundred yards—with a good cooking coal.

Previous to 1860, it was the custom of the Kimbrough Bros. to mine coal from the seam at Rockwood for the use of blacksmiths for miles around, and the same coal had been opened at various other points, among them a property north of the Emery River. This property had formerly been worked for blacksmith use by DeArmond, as the seam at Rockwood by Kimbrough. In 1866, W. and E. Small, of Baltimore, purchased 1,200 acres of land on Walden's Ridge from John DeArmond, and commenced mining coal. They failed in 1869, and in 1870 the property was purchased by the Wilcox Brothers, who mined coal there a few years and also failed. The mine was worked by a tunnel, which, going in horizontally, struck the seam 210 feet below the outcrop; entries were driven to the right and left, and coal mined by stoping from below up. The seam is very regularly five feet thick. The cut on page 16 was made by Prof. Bradley to illustrate the location of the strata at this mine, from it may also be derived a very good idea of the whole of Walden's Ridge uplifted strata. Recent examinations which I have made fully satisfy me that these Walden's Ridge seams go down under, form a basin and become for a time horizontal, and then rise again at the line of the Sequatchie fault. This fact is well illustrated along the line of the Tennessee and Sequatchie Valley Railroad, near Spring City. By following this line to Swaggerty's Cove, a very fine section illustrating the peculiar location of the strata can be obtained. As the road climbs Walden's Ridge the inclined strata with the pitched seams of coal are plainly seen, then on the mountain the road descends to the valley of White's Creek on barren strata, there reaches the coals again and ascends with them to Jewett, where are the mines of the Spring City Coal and Coke Company; from Jewett to the summit of the mountain forming the eastern rim of Swaggerty's Cove, the rapid rise of the strata is very plain, the coal seams, where worked, having a rise to the north-northwest of six feet to the one hundred. The Seral conglomerate forms the top of the mountain immediately bounding the eastern side of the cove, and the mountain limestone comes in about two hundred feet below, being there higher than it is at the foot of Walden's Ridge on the east. In the valley of White's Creek, the basin formed is well illustrated by a coal seam owned by Messrs. J. C. Wasson and T. J. Neal, which is so nearly level as to be difficult of drainage. It is an excellent coal, however, and very convenient to the railroad. My opinion is that it is one of the Lower coals, probably the equivalent of the old Etna seam, at this point, however, appearing to be very regular.

The Emery mines are not now worked, and have not been for several years. It is a valuable property, and the coal made a strong good coke, nearly free from sulphur. In the neighborhood is a good site for an iron furnace.

At Emery Gap, Messrs. Byrd and Denning have opened the main or Rockwood seam, and have mined considerable coal for local use and some for shipment. I may here remark that this seam is classed by Prof. Bradley as No. 6, while at Soddy it is called No. 7, and at Etna its equivalent is called No. 6. It is probable that the Soddy section is not correct. In Safford's Sewanee section it is classed as No. 5.

THE ROANE IRON COMPANY'S MINES

are located at Rockwood, Tenn. The seam of coal at this place was first opened for regular operations in 1867, by Gen. J. T. Wilder and Maj. H. S. Chamberlain, who conceived the idea of erecting an iron furnace. The seam was originally opened in a gulch where it outcropped in the end of the ridge, various other openings were afterwards made, and the freaks which nature has played with this seam of coal at this place probably have no equal anywhere. At one point the seam widens rapidly until it reaches a thickness of one hundred and twenty feet, then at other places it thins down for several hundred feet to a mere thread. A peculiarity of this seam of coal is a strip of putty-like clay, which never ceases to exist even in the places where the coal is perfectly squeezed out. This seam of putty-like clay is common to the Sewanee seam also, as I have lately determined by an examination at Tracy City. An idea of the manner in which this mine was first worked may be formed from the following diagram:

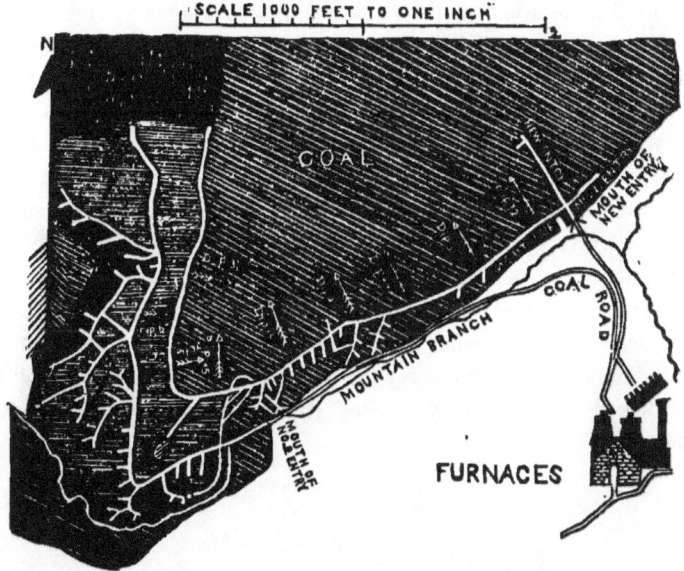

The original point of opening is seen to the left, while the tunnel from which the chief supply is now drawn is on the extreme right. The usual dip of the coal is from 30 to 35 degrees. This diagram was made by Col. Killebrew in 1877, and while it serve the design of showing how the coal has been mined, does not now truly exhibit the present interior shape of the mine. The arrows indicate the dip of the coal.

In 1876 this company determined to try a new plan, and struck boldly into the mountain with a tunnel from near the valley level. The coal was reached at a distance of 500 feet, about 200 feet vertical height below the outcrop, and 300 on the line of the dip. From this tunnel entries have been driven to the right and left for considerable distance, and ventilating slopes connected to the openings above. On the right hand entry about 150 feet from the tunnel a slope has been driven down for 365 feet, at which point coal seam becomes very nearly horizontal. A steam engine with drum and wire rope lowers and raises the cars. The slope is ventilated by a parallel slope or entry. This work demonstrates practically that the seams of the inclined strata of Walden's Ridge go down and under the mountain, thus proving the existence of an enormous amount of this excellent furnace fuel. The location of this tunnel is demonstrated by the following diagram, and from its success is proven that the same species of operation may be conducted a large part of the distance from Dayton to Cumberland Gap. It is proper to state that the Company very prudently ran a slope down on the dip of the seam before commencing the tunnel; The tunnel is nearly all the way through sandstone, and cost about $7,000. The mouth of the tunnel is two hundred feet above the level of the Cincinnati Southern, hence the point at which the coal seam becomes horizontal is only one hundred and fifty-seven feet lower than the railroad bed. The rocks in the valley, where the railroad runs, are pitched at fully as great an angle as the coal. The shale shown at the right hand end of the diagram has been largely cut away thus shortening the actual tunnel length.

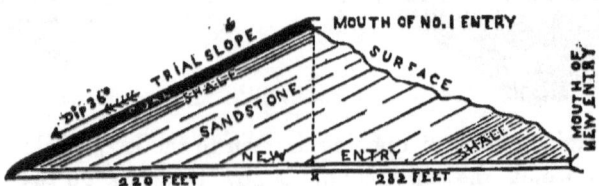

The coal from the Roane Iron Company's mine is used entirely in their furnaces, except a little for domestic use of their employes. Their furnace plant consists of 120 coke ovens and two stacks, which when both are running, turn out an averrge of 115 tons of pig iron per day. From a wild forest in 1868, this place has now grown to a thriving town with a population of about 3,000 inhabitants.

During the year 1882 this company mined and made into coke 1,500,000 bushels of coal. The average number of hands employed in the mines was one hundred and seventy-five, whose usual earnings were from $2.00 to $3.00 per day, and to whom $92,972.78 were paid in wages. The value of the coal used is estimated at $103,500. The total paid for mine supplies during the year was $23,511.33. During 1882 the new slope was sunk, hoist engine put in, new air shaft made and 8,000 feet of new iron track put in the mine. Two cents per bushel was paid for mining, $2.25 for driving entries in coal, in slate and rock $10.00 per yard. There are three main entries, one 1,200 yards long, another 1,700 yards and another 1½ miles, all 7x6 feet in size; the side entries are from 7 to 10 feet wide and 6 feet high. Work was carried on every day of the year. There are good schools with separate buildings for white and colored children, and four churches. The cost of living was a little higher than in 1881. Reuben Street is Superintendent of the mines, H. S. Chamberlain is President and H. Clay Evans General Manager of the company.

At the gap of the ridge through which White's Creek passes, some openings have been made but no mining operations for shipment conducted. The coal there is of good quality, and the iron ore close at hand and in abundance, there is a good location for an iron furnace.

THE SPRING CITY COAL AND COKE COMPANY.

The mines of this company are not properly in the inclined strata of Walden's Ridge as I have classed them, but as they are reached from the main front and over the inclined strata, they will be here connected therewith. On a previous page the peculiar location of these mines has been alluded to and the excellent section which can be had on the railroad line from the valley to Jewett, where they are located. The mines are sixteen miles from Spring City, and about 1,000 feet elevation above that place. The openings have been made in the ridges on each side of a ravine which appears to be also the line of a nearly east and west fault. On the south side of the ravine three openings have been made and the seam ranges in thickness from one to three and sometimes four feet. It dips to the east at the rate of six feet to the one hundred. On the north side two narrow seams were found separated by from five to twelve feet of slate. An entry has been driven in 1,400 feet and the seams have come within a few feet of each other, and give every appearance of uniting. It is my opinion that this seam is the Rockwood seam both from the peculiarity of the coal as well as the fossil plants of the roof, and if not already united when these lines are printed, it is my opinion that pushing the entry under the main and undisturbed mountain will find them united in a seam of excellent coal. The buildings, tracks and general plant of these mines are well constructed, but it would be difficult to pick out a poorer location for opening a mine, yet it may be more economical to retain it as there than to move. The business seems to be well managed now.

As stated this seam dips at the rate of six feet in one hundred. I do not think this continues very far as the basin where the rocks lie perfectly horizontal, is not over two miles distant to the east and I think one and a half will cover it. This point is 300 feet lower than the mine opening, and there the excellent block coal noted previously is found in the bottom of the creek about fifty feet lower, having above it what is apparently the conglomerate. In the west the strata rise still more rapidly, and I have no doubt but the outcrop of the seam they work could be found on top of the narrow plateau which there bounds Swaggerty's Cove, and the Lower Measures in their place under the conglomerate which forms the wall-like sides of the cove. As previously stated this is a very interesting region for the student of geology. Regular operations in and shipments from this mine were commenced in the summer of 1882, hence the manager could not make a full report. About 8,000 tons of coal were mined and shipped in 1882. The company has expended about $50,000 for improvements and labor. The railroad is owned distinctively from the local mines; usual freight to Spring City $4.00 per car, freight from Spring City to Chattanooga 50c per ton. Col. Chas. Clenton is General Superintendent and also President of the Railroad Company, and Wm. Sleep,

Mine Manager. Mine post-office, Jewett, Cumberland county; company office at Spring City, Rhea county.

In the inclined strata of Walden's Ridge, near Grand View, coal has been mined in a small way and the quality is said to be good. The outcrops of these seams are plainly visible from the windows of the railroad cars while ascending the mountain, and there is no doubt that the Rockwood seam could be easily found. The elevation of Spring City is 767 feet above the sea, Grand View is about 850 feet higher, and the outcrop of what is thought to be the Rockwood seam has an elevation of about 750 feet above Spring City, or 1,537 feet above the sea, and its dip is over 45 degrees. From Grand View there is a steady down grade to the west, probably to an elevation of over 1,300 feet, then there is a rise at Jewett to 1,767. Hence the coal strata has somewhat the following shape:

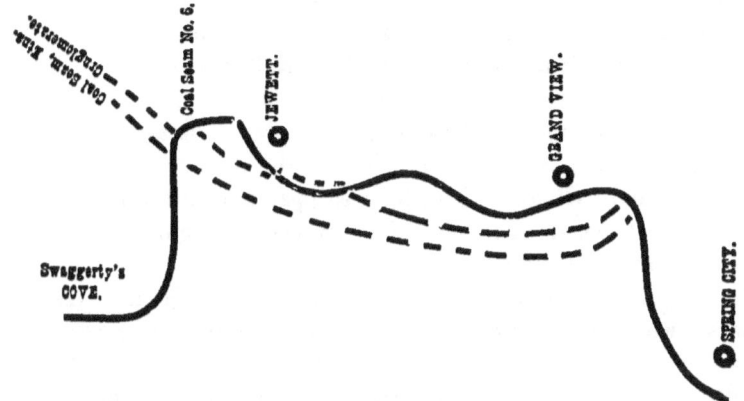

This section is not intended to be on any scale, but simply to give an idea of how the coal appears on the line of the Sequatchie Valley and Tennessee River Railroad. From Grand View to the rim of Swaggerty's Cove is ten miles, while from top of rim to level of Cove is only about 300 feet. The Clinton shales appear in Swaggerty's Cove, and, I was told, also, the iron ore, but did not see it. The Nashville limestones are also said to appear.

It is a remarkable fact that the coal seams in the lower part of the Cahawba field have a similar formation. The sharp upturned outcrop being near Montevallo, and the long sloping arm reaching out towards Shades Creek.

During my visit to this region limited time did not permit me to make a measured section of the line indicated in the diagram. There can be no doubt that such would be of great value in a proper study of the geology of the State.

Near Clear Creek, eight miles southeast of Spring City, Mr. Wm. Stambaugh has opened a number of coal seams in the inclined strata of Walden's Ridge on an area of land belonging to Mr. John Stambaugh, of Youngstown, Ohio, and has developed the Rockwood seam.

DAYTON COAL AND IRON COMPANY.

The mines of the Dayton Coal and Iron Company are in the gorge of Richland Creek, at the passage of which the Walden's Ridge inclined strata commence to break down. The mines opened however, are in a horrizontal strata very high above the level of the creek, and some distance west of the line of the inclined strata. The position of the seams they have opened has not been fully determined, but it is my opinion that they are 9 and 10 of the Soddy section, 8 and 9 of the Daisy section and the Kelly and slate seam of Etna. This opinion, however, is not based on as thorough an investigation as I would desire. So far the seams developed have been thin and variable. They have shipped but little coal, making it mostly into coke. The lands were originally bought with a view to the erection of a blast furnace, and this action is not yet abandoned. The property is owned by Sir Titus Salt, of England. He also owns a wide gauge railroad to the Cincinnati Southern, which has been graded on to the Tennessee River. The entire plant is well and substantially constructed. The railroad was only completed so that coal could be shipped in December, 1882.

The property consists of 38,000 acres of land. Only fifty tons of coal were shipped before the end of 1882. The average number of hands worked during that year was sixty, to whom $1.00 to $2.00 per day was paid, amounting to a total of $45,000, the sum of $26,000 was paid out for all supplies. Miners are paid $3\frac{1}{2}$ cents per bushel, $4.50 per yard for main entries and $4.50 for cross entries. The main entry is 700 feet long, 6x8 feet, side entry 300 feet long, 5x8 feet. Three schools are taught in the neighborhood, and there are two churches. Freight to Chattanooga was 58 cents per ton. John H. Ferguson, Dayton, is General Superintendent.

HORIZONTAL STRATA.

A few miles below Dayton, the last remnant of the pitched strata of the Walden's Ridge sinks to a level with the valley formation. From thence southward there is no distinction in the mountain formation, and though the coal seams may be slightly inclined at their outcrop, this soon ceases, and a horizontal basin is reached, which, though not yet fully demonstrated, probably rises again to the northwest. This seems to be proven by the position of the mountain limestone, which at Soddy, Daisy, and other points north of where the Tennessee River cuts through the ridge, is down in the valley only about 800 feet above sea level, and at Etna is also but little on the mountain side, is on the northwest escarpment above Sequatchie Valley more than 1,100 feet in elevation.

This point has also been partially demonstrated at the Soddy mines. Their main entry is run into the mountain at right angles to the general trend of the Walden's Ridge, that is a line about north 75 west. It is now 1,000 yards long; for the first 150 yards it dips into the mountain at the rate of one foot in thirty feet, there is a level for a short distance, then comes a rise, the incline of which is somewhat less than the dip named. Of course a full demonstration of this theory cannot be had until the Soddy Company go through the mountain, but there is sufficient evidence to base an assumption of its correctness.

As previously stated in this book, the Raccoon Mountain being a mere continuation of the mountain called Walden's Ridge, northwest of Chattanooga, only separated by the river, therefore it would be treated as the same and its coal seams considered with those of Walden's Ridge. Hence I insert here three sections, showing the very great similarity of the strata at three different points. The Etna section is from Dr. Safford, the Daisy I made, and that at Soddy is from Capt. A. Lloyd, Superintendent. He appears to have one more seam of coal than at the other places, which I am satisfied is a mistake, coal 0 in the gray shale being undoubtedly a slip and should not be counted, and I have changed his numbering. No. 10 of the Etna and Daisy section was called No. 11 at Soddy, but has never been worked. It is on top of the mountain, and near the brow has very little covering on it. It has been opened at Daisy and two entries driven into it, and the evidence that it is identical with No. 10, or the Oak Hill seam at Etna, is positive, as also is the identity between the Kelly seam of Etna and No. 8 at Soddy. This seam has not been sufficiently opened at Daisy to determine its quality accurately, but its position is correct. It is my opinion that the Soddy main, or No. 6 seam, is No. 6 at Etna; at Daisy it is known to exist, but its thickness has not been determined. At Etna it has been slightly opened, and has increased from one foot in thickness to nearly two feet. Some experimental openings have been made by C. E. James & Co., on the Walden's Ridge, across the river from Chattanooga, at which several seams of coal have been developed, but no one could give me an idea of the section, and time did not permit to visit them. I see no reason why coals No. 10 of Etna and Daisy should not exist there. A coal seam has also been opened near the suck by the McNab Coal Company, which is thought to be No. 6, or the Soddy seam, but it is probable that it has not been sufficiently examined or developed to establish its identity. A narrow-gauge railroad, called the Chattanooga Western, is nearly finished to the mines of C. E. James & Co.

The following are the sections spoken of:

SODDY.

Total elevation 1,980 feet.

10.	Coal top of mountain	4		feet.
	Sandstone	83		"
	Shale	1½		"
9.	Coal	2¼		"
	Gray shale	40		"
8.	Coal	2 to	4	"
	Fire clay	6		"
	Sandy shale	40		"
7.	Coal	1½		"
	Sandy shale	40		"
6.	Coal	3 to	7	"
	Fire clay	2		"
	Shales with thin coal			
	Sandstone	100		"
5.	Coal	2		"
	Fire clay	2		"

	Black shale	40		feet.
4.	Coal	1¼		"
3.	Gray shale and thin coal	35		"
2.	Coal	2½		"
	Fire clay	8		"
	Shaly sandstone	40		"
1.	Coal	1¾		"
	Sandstone and shales	68		"
	Mountain limestone			

DAISY.

Sandstone heavy is 4,000 feet back from the cliff, as also at Soddy, is called the second bluff.

11.	Coal	1	to	1½	feet.
	Shales	20			"
	Sandstones called first bluff, about 1,000 feet from cliff	60			"
10.	Coal good	4	to	5	"
	Fire clay and shales, with thin coal	40			"
	Sandstone, heavy, massive, called main bluff	50			"
9.	Coal	2½	to	4	"
	Shales and black slate	30			"
8.	Coal	3			"
	Fire clay	3			"
	Sandstone in layers	20	to	30	"
7.	Coal, thin, and shales	50			"
6.	Coal, estimated	2	to	5½	"
	Shales	10	to	20	"
	Conglomerate cliff rock	90			"
5.	Coal (Etna)	1	to	5	"
	Fire clay	1	to	4	"
	Sandstones and sandy shale	3			"
4.	Coal, thin	½			"
	Shales with iron balls	80			"
3.	Coal	1			"
	Shales	15			"
2.	Coal	2½	to	3	"
1.	Coal, thin, and shales	100	to	125	"
	Mountain limestone				

ETNA.

	Sandstone	73			feet.
	Shaly sandstone	32			"
10.	Coal	4			"
	Slate with a thin coal	46			"
9.	Coal with slate parting	6			"
	Shale	44			"
8.	Coal (Kelly seam)	2	to	5	"
	Fire clay	1½			"
	Sandstone (upper conglomerate)	82			"
7.	Coal	¼			"
	Yellow sandy shale	45			"
	Thin coal				
	Gray shale	47			"
6.	Coal	1	to	2	"
	Gray sandy shale	45			"
	Conglomerate cliff rock	96			"
5.	Coal main (Etna)	2	to	5	"
	Fire clay	2			"
	Shale	20			"
4.	Coal	1			"
	Gray shale	95			"
3.	Coal	½			"
	Black shale with iron	20			"
2.	Coal	3			"
	Shale with iron balls	40			"
	Gray shale	34			"
1.	Coal	2			"
	Fire clay	2			"
	Shales and sandstone	100			"
	Mountain limestone				

THE SODDY COAL COMPANY'S MINES

Are located on the Cincinnati Southern Railway, twenty-one miles from Chattanooga, at Rathburn Station. The land now worked is leased from Wm. and M. H. Clift, the royalty paid being one fourth of a cent per bushel, equivalent to 6¼ cents per ton. For the year ending December 31, 1882, 2,500,000 bushels of coal were shipped to Chattanooga, Atlanta and Macon, Ga., Huntsville,

Ala., and other southern places. During 1882, 700,000 bushels of coke were made and shipped. The average number of hands worked was 300, and their usual earnings from $1 to $3 per day. The average selling price of coal for the year was $6\frac{1}{2}$ cents, and the value of coal shipped amounted to $165,000. For driving entries, $2.35 to $4.50 per yard is paid, and mining coal is paid for on a sliding scale as follows: The miners are paid $67\frac{1}{2}$ cents per ton when coal is sold at 1.87\frac{1}{2}$ or less, and they receive one-third of any amount coal is sold at over 1.87\frac{1}{2}$ per ton; for instance, if the average of monthly sales is $2.00, that is $12\frac{1}{2}$ cents more, the miner receives four cents per ton more, or $81\frac{1}{2}$ cents. The main entry is 1,000 yards long, $5\frac{1}{2}$ by 8 feet in size; side entries are 5 by $7\frac{1}{2}$. Rooms are worked with a width of twenty four yards and a roadway on each side; pillars are left twenty yards wide. In 1882 300 days of full time were made. Liquor is not allowed to be sold near the mines. There are three churches and good schools. Cost of living was a little higher than in 1872.

This mine was opened in 1867 by an association of Welchmen on the co-operative plan. It proved a failure, and the mine went into the hands of a Receiver. The present company took charge in 1877, and the business has steadily increased, the capacity having more than doubled in the last two years. Cost of shipping to Chattanooga is $2\frac{1}{4}$ cents per ton per mile by railroad, and less than one cent by water. Capt. A. Lloyd is Superintendent at the mines, M. H. Clift, President, and Wm. Williams, Agent in Chattanooga.

Three seams have been worked at this place, classed in the section as No. 6, No. 8 and No. 9. The chief mining operations are in the first. Near the outcrop it was somewhat friable, but farther in it became solid, and is now a coal of very good quality and much sought for. It is so compact that lumps of great size can be easily taken out.

This seam is worked by four entries, the largest of which is over 1,000 yards in length. The seam rises to the southwest, up the creek, about 75 feet in a little less than three-fourths of a mile. At that point an entry has been driven and a ventilating shaft put in, hence there is a fine current of air up and a constant stream of water down. It would be hardly possible to effect a more perfect system of drainage and ventilation than has been effected through the opportunity afforded by nature.

No. 8 is a seam of extraordinarily good coal, ranging from twenty inches to four feet thick. Alone it makes a very superior coke, and is no doubt the same as the Kelly coal at Etna. It is usually mixed with the seam below in making coke, thereby improving the quality. This mine is worked by contract, the company only furnishing props and trackway, and the "boss miner" delivering the coal in the shute at three cents per bushel.

No. 9 is but little worked of late, as its quality is inferior.

From the opening of these mines to the 1st of January, 1883, about 600,000 tons of coal have been taken out, and the annual output is now about 140,000 tons.

The mine incline is 175 yards long, about 25 feet vertical height, and from the foot of the incline to the Cincinnati Southern is one mile. The entries have a total length of 9,000 yards. In summer the men are chiefly employed in driving entries, so as to prepare for a large output in winter, and May, 1883, they drove 225 yards of main entry and 435 yards of room entries. The main entries are all driven with a parallel side entry for ventilation. The pillars are left twenty yards wide. In operating the No. 6 mine, two faults have been encountered which nearly cut out the coal; in No. 7 only one has been met so far, but that seam has not been run in near as far as No. 6.

THE WALDEN'S RIDGE COAL COMPANY

Is the title of a corporation with the same stockholders as the Soddy Company. It operates a mine on Rocky Creek, nine miles farther up the railroad. Operations were commenced there since January 1, 1883. The product is now (June, 1883), about 75 tons per day, all of which is made into coke. About forty men are employed in the various operations of mining, coking and construction.

THE DAISY COAL COMPANY

Is the title of a corporation owning about 8 000 acres of land adjoining the Soddy property on the southwest. Four miles from Soddy, and in a total distance of 6,000 feet of the railroad, a number of openings have been made. These are in Nos. 3, 5, 8, 9 and 10 seams. The principal work has been done in No. 5, which is the same as the old Etna, and has all its characteristic fickleness as to thickness, but is a very superior coal. In No. 10 an entry 110 feet long has been driven, revealing a seam of good coal 4 feet 10 inches thick. This seam is on top of the mountain, in a sort of bench, which is very persistent, running about 1,000 feet from the front bluff of the mountain, from the Soddy mines to Chicamauga Creek. I am not informed as to its existence beyond Chicamauga. Operations in these mines have been of the most desultory character for some time, the present owners not being able or willing to make the further investment necessary to construct the railroad and shutes.

An estimate made by a mining "boss," as the cost of mining and shipping 5,000 bushels per day, is as follows, and may fairly apply to other places in this section:

10 Mules...$	5 00
10 Drivers..	10 00
2 Weighers and dumpers.....................	4 00
4 Mules to incline................................	2 00
4 Men to incline...................................	4 00
1 Incline man.......................................	2 00
1 Man at bottom..................................	1 00
Engineer and fireman on railroad........	5 00
Brakeman...	2 00
Contingencies.......................................	5 00
	$ 40 00
Mining 5,000 bushels @ 2½c................	125 00
Props. " " ½c................	25 00
5,000 bushels at C. S. R. R..................	$190 00

This estimate was made for the No. 10 seam, wherefrom there would have to be a tramway to the incline of 1,100 feet in length.

This is a very valuable property, and it is to be regretted that it is permitted to lie idle. The distance to Chattanooga is seventeen miles. J. C. Gaut, Jr., Nashville, is President, and Capt. Bueckell has had the mines in charge, and has been getting out a little coal for sale to the neighborhood and to Chattanooga.

It is a matter worthy of note that seam No. 7 is 1,221 feet above sea level at the main entry at Soddy, while at Daisy it is 1,314, and the seam supposed to be the same at Etna is 1,542 feet above sea level.

In the Walden's Ridge, nearly a direct north course from Chattanooga, C. E. James & Co. have opened several seams of coal.

A company has been organized, under the name of the Chattanooga Western, Maj. G. C. Connor, President, to construct a railroad from the city to these mines, and some of the most difficult parts of the route have been graded. They probably have all the seams which exist at Daisy and Etna.

[Since the above was put in type, the Daisy property has been purchased by Messrs. J. G. Aydelott, French and others in company with Mr. Thos. Parkes; these gentlemen state that they will immediately commence operations on an extensive scale.]

Lower down the river another coal operation was commenced and prosecuted with some vigor for a time, under the name of the McNabb Coal & Coke Company.

THE ETNA MINES

Are located near the Nashville & Chattanooga Railroad, fourteen miles from Chattanooga. They are located in the continuation of the mountain which, northwest of Chattanooga is named Walden's Ridge, here locally called Raccoon Mountain, and, as previously stated, a little further southwest is called Sand Mountain. The location of these mines is probably the best to study the development and characteristics of the Lower Coal Measures. A section of the strata is given on page 68.

A railroad track, 7,200 feet long, extends from the Nashville & Chattanooga Railroad to the top of the plateau. This line skirts the mountain side, cutting away the earth and rock, and then through several long deep cuts it reaches the top. These cuts afford excellent opportunity to study the various formations. In the railroad cut near Whiteside a thin seam of red fossil iron ore shows; further east several seams over a foot thick are found. Above this formation is found the siliceous group and the mountain limestone. The mountain limestone shows just below the line of the Etna Mine Railroad, the first formation seen above being the shales and shaly sandstones of the Lower Coal Measures. In these is found a thin seam of coal, above this is a wide belt of shales containing modules of carbonate of iron; then comes a seam of coal which varies from two to six feet in thickness. It has been worked to a limited extent, and is a good coking coal, but contains considerable slate. It is thought to be the same seam as that worked at the Dade mine in Georgia. In the black slate above this coal is a very persistent seam of iron ore about twelve inches thick, which can be easily mined with the coal, especially in entries, and from which could thus be made available a considerable quantity of ore. Between this seam and that called the main Etna are two thin seams of coal of no importance. The main Etna was the seam chiefly worked for many years; it is now abandoned. It was an excellent coal, but the seam was so variable in thickness as to make mining very costly.

It is the same seam as that found immediately under the cliff conglomerate, all around the Highland rim, from the tunnel, through White, Putnam and Watson counties, into Kentucky. At the Etna mines there are several seams between the cliff and the second conglomerate; one of these is undoubtedly the Sewanee. In a former work on "Coal and Iron in Tennessee," the writer of this, who also prepared that part of that work, expressed the opinion that the Kelly seam was the Sewanee; later and more careful investigations prove that statement to be incorrect. The Sewanee seam is here thin, as far as examined not over twenty-four inches, at the outcrop only thirteen inches thick. Above the second conglomerate is a plateau over which is a railroad one thousand feet long to reach a superimposed ridge. In this ridge are three seams of coal. The lowest is called the Kelly, the next the slate and the top Oak Hill. The first and the last are worked. The Kelly coal has probably no superior anywhere as a coal for blacksmiths' use or for making coke for the use of founderies. It very much resembles the famous Blossburg coal of Pennsylvania. This seam dips to the northwest, this for some time gave great trouble as to water, which was allowed to run to a swamp, and was pumped thence by an engine located near the mouth of the mine. Within the past two years, however, entries have been driven through the ridge at two points, which not only give excellent ventilation, but by a side ditch afford egress for the water and save the cost of pumping. The seam ranges in thickness from eighteen inches to four feet; the cars used in the mine hold sixteen bushels; they are carried to the mine mouth by mules, thence they are hauled up an incline by a wire rope worked with a steam engine, then they are dumped into cars holding ninety-six bushels, or six small car loads. These large cars go down the incline to the coke ovens and railroad.

The Oak Hill seam was opened many years ago and for some

time abandoned, but has again been operated for two years past; the coal from it is cubical and laminated, and makes a good, strong coke, but contains considerable sulphur. If washed it would undoubtedly be of first quality. It is very easily mined, though the roof is peculiar and somewhat unreliable, and it may certainly be counted on for four and a half to five feet of coal. Great care in mining might get out a large part of the sulphur, as it is generally in plates near the top. It is at present impossible to identify this seam of coal with any in other parts of the Tennessee coal field, except at Daisy. It does not exist at Tracy City, nor so far as known anywhere further north, though I am inclined to think an outcrop I found on a high ridge near the headwaters of Rocky River, is this seam. At any rate that outcrop is there high enough above the Sewanee to be in the proper place; on the eastern side it has not been observed northeast of Soddy. The Slate seam carries a coal very free from sulphur, but containing a large amount of slate disseminated through it.

There was shipped from the Etna mines during the year 1882, 22,730 tons of coal and 7,609 tons of coke; the total number of employes was 190, of whom eighty-four were miners and ten coke burners; the total amount of wages paid out was $63,510; men working five days to the week earned $600 per year; common labor is paid $1.00 to $1.25 per day.

The postoffice of the mine is Whiteside, in Marion county; they were originally opened about 1852, by an Eastern company working under a lease from Robert Cravens, and the Boyce and Whiteside estates. Since that time they have been operated by several different companies and individuals with varied success and reverses. The present company was organized in August, 1881, under the name and style of the Etna Coal Company, with the following officers: Dr. Wm. Morrow, President, D. B. Pillsburry, General Manager, and J. T. Hill, Secretary, which organization still exists. The mines now operated are owned by the company, the estate consisting of about three thousand acres, extending from the Nashville, Chattanooga & St. Louis Railway on the south to the Tennessee River on the north. At present the only seams worked are the "Kelly" and "Oak Hill," the former being the principal seam, and the one from which the celebrated blacksmith coal is obtained. "This coal has been in general use by blacksmiths throughout the Southern States for more than twenty five years, and is now fully established as the best blacksmith coal in the South, and equal if not superior to any coal for that use in the United States." This mine was originally opened for general domestic use and the product was sold largely in Nashville, Chattanooga, and elsewhere, but its superior qualities for blacksmith use and for the manufacture of coke soon caused the trade to drift almost exclusively into that channel. At present almost the entire output is sold to blacksmiths throughout the South.

The following analysis of the Etna blacksmith coal was made by Prof. Pohle, of New York City:

Fixed carbon..74.20
Volatile matter...21.10
Ash.. 2.70
Sulphur... 70
Water... 1.30

A recent analysis made by Prof. Andrew S. McGrath, of Pennsylvania, to determine the amount of phosphorus in this coal and coke, shows the following result:

Coal..0.005 per cent.
Coke..0.008 per cent.

Being 0.006 per cent. less than is found in the famous Connellsville coal, so highly esteemed in the manufacture of Bessemer pig iron on account of the low per cent. of phosphorus.

The product from the "Oak Hill" mine is manufactured into blast furnace coke of good quality, which finds a ready sale.

MIDDLE TENNESSEE.

SEWANEE DISTRICT.

In this district is included the Tracy City mines and the mines in the Sequatchie Valley, also the Crow Creek region. The principal seam worked is the Sewanee; in fact, in it are the only extensive mines. There are several openings in the mountain above Crow Creek Valley, but they are entirely in the Sub-conglomerate Measures. The peculiarities and excellencies of the Sewanee seam have been stated in another part of this volume. It is undoubtedly the most extensively distributed seam of ary in the Tennessee coal field. In the Cumberland Plateau region it underlies about one-half of Grundy County, one third of Sequatchie, full three-fourths of Van Buren, one half of Bledsoe, over one-half of Cumberland, and some area in the southern part of Fentress and western part of Morgan—an area of about 1,300 square miles in this district alone. The seam is undoubtedly the same as that designated B, by Prof. Lesley in the Geology of Pennsylvania. It occurs in Pennsylvania just as it does in this State, above the main (Seral) Conglomerate, at a slightly variable distance from it. In that State there is usually, but not invariably, a coal seam below. This seam is found at the Tracy City mines. This seam B is but little less important as a coking coal than the Pittsburg seam in the Connellsville region; it is the same seam from which large quantities of coke are made in Pennsylvania. The Pittsburg seam, from which the largest amount of coke is made, in the Connellsville region, has a composition very near identical with Sewanee, but farther west in Westmoreland and Alleghany counties, the volatile matter is higher and the carbon less. For instance:

	Fayette County. Frick & Co., Broad Ford.	Westmoreland. Pen. Co.'s Mines.
Water	1.26	1.490
Volatile matter	30.107	37.153
Fixed carbon	59.616	58.193
Sulphur	.784	.658
Ash	8.233	2.506
	100.00	100.00
Color of ash	Reddish gray.	Red.

The two coals given above, as stated, are both from the seam known as the Pittsburg seam, and are from mines not thirty miles apart.

Two analyses of the Sewanee coal from Tracy City mines are:

Water	1.77	......
Volatile matter	25.41	29.9
Fixed carbon	62.00	63.5
Ash	10.82	6.6
Sulphur		Trace.
Fuel ratio	1:2.43	1:2.12

The coals of seam B in Pennsylvania have less volatile matter than the Sewanee. This is attributed to their proximity to the anthracite fields, it having been ascertained that the volatile matter increases as the seams go westward to the Ohio. The mines of the Cambria Iron Company, in Blair County, bear closest resemblance to Sewanee of any of the B coals of Pennsylvania.

Two of the analyses are:

Water	.950	1.400
Volatile matter	28.915	27.235
Fixed carbon	63.462	61.843
Ash	5.690	6.930
Sulphur	.983	2.602
Fuel ratio	1:2.27	1:2.19
Color of ash	Gray.	Gray.

These and other Pennsylvania analyses are copied from the report of Prof. A. S. McCreath, Chemist of the Geological Survey of Pennsylvania.

It is seen from these analyses and from the area covered by this seam, as stated in another part of this book, that in this coal we have a vast treasure of wealth stored up for future use. On the southeastern side of the mountain it is very convenient to Chattanooga; it is also convenient to Sequatchie Valley, where there is now partial transportation, and in Middle Tennessee it may be reached at various points by branch lines from the McMinnville branch of the Nashville, Chattanooga & St. Louis Railroad; the seam appearing in the hills of the plateau for some distance parallel with that line, within a distance of fifteen to twenty miles. Thus is afforded a storehouse of fuel for the conversion of the vast amount of iron ores of the Western Belt into metal, and giving a possibility for the erection of many furnaces.

The chief mining operations of this district are those of the Tennessee Coal, Iron & Railroad Company, who own a large area of land.

THE TRACY CITY MINES.

are the largest operations of this company, and also the largest single mining operation in the State. Its history is an interesting commentary on the progress of coal mining in the South, it being the pioneer enterprise of any size in this section of the Southern States.

The seam of coal at this place was discovered by some boys hunting a rabbit; the animal ran under the root of a tree which had been blown down; in digging it out the coal was found. They reported the discovery to their father, Ben Wooten, and he, thinking it might be of some value, got out a grant for 500 acres, covering the opening. The Wooten Brothers afterwards opened the seam, and for many years hauled the coal down the mountain to the blacksmiths in the valley, and some was sent to Nashville. In 1852 Messrs. Boorman Johnson, John Cryder, S. F. Tracy and other gentlemen of the city of New York, came to Tennessee looking for opportunities for investment. They were shown the Wooten coal bank, and immediately opened negotiations for the property and soon after purchased the land. A company was then formed under the name of the Sewanee Mining Company, which had a cash paid in capital of $400,000. In 1854 the construction of a railroad from the Nashville & Chattanooga Railroad to the mines was commenced. Even at this day the building of a railroad up that mountain would be thought a stupendous undertaking. The grade rises 736 feet in 5.87 miles, and the curvature is very great. This road was finished in 1859, but left the company in debt in the sum of $400,000. The company was sued both by New York and Tennessee creditors. The Tennessee creditors, represented by Hon. A. S. Colyar, obtained the first judgment, bought in the property and reorganized the company, under the name of the Tennessee Coal & Railroad Company. This company commenced operations in 1860, with Hon. A. S. Colyar as President. In 1862 the mines were abandoned by the company, but were taken possession of by the United States troops and worked for the use of the army for some time. In 1865 Col. Colyar effected a compromise with the New York creditors, giving them bonds and some stock, and, with Col. P. A. Marbury as General Manager, recommenced operations. Later, Mr. J. C. Warner became General Manager, but afterwards took the place of Secretary and Treasurer, and Mr. A. M. Shook was elected General Manager. The history of the early days of this enterprise is a grand record of energy and perseverance, and, as usual, these great qualities triumphed, and the largest coal operation of the State was put on a sound footing.

To these gentlemen is largely due the present extensive coke industry of the State. In 1873 they foresaw that to make a great and profitable business the manufacture of coke must be entered into, and that that coke must be a good iron-making fuel. A small furnace was erected on the mountain and this experiment satisfactorily tested. This was in 1873, but previously thereto coke had been made in pits on the ground. The first coke was shipped in 1868. In that year 5,377 bushels were shipped. The further growth of this business can be gathered from the tables below. The great impetus was given to it by the erection of the Chattanooga Iron Company's blast furnace in 1873. As will be seen from the table, the coke shipments jumped from 62,175 bushels in 1873 to 619,403 in 1874. In 1875 the entire property was sold to Cherry, O'Connor & Co., A. M. Shook retaining an interest and being General Manager. In 1880 these gentlemen commenced the erection of the furnace at Cowan, which was finished in July, 1881. And in the early part of 1882 they sold the whole property to John H. Inman, Smith and others, Tennessee parties retaining a one third interest. The name was then changed to the Tennessee Coal, Iron & Railroad Company. The first coal shipped from this mine since the war was in June,. 1866. During that month 680 tons were shipped, and for the seven months of that year the shipments were 9,240 tons; from that time to January, 1883, the shipments have been :

Year.	Coal—tons.	Coke—tons.
1866	9,240	
1867	36,250	
1868	40,850	107
1869	36,880.	413
1870	47,110	673
1871	61.940	668
1872	102,320	455
1873	99,380	1,243
1874	93,370	12,388
1875	109,100	16,160
1876	109,689	21.080
1877	103,951	21,060
1878	89,000	22,780
1879	93,037	44.800
1880	114,170	64 440
1881	158,679	85,022
1882	144,689	108,153

The coal is calculated in tons of 2,000 pounds, 80 pounds to the bushel; the coke in tons of 2,000 pounds, 40 pounds to the bushel and 50 bushels to the ton.

The company own in this their original property about 25,000 acres of land, of which it is estimated that 10,000 are underlaid with the Sewanee seam of coal; with lands since purchased adjoining this tract, it is probable that they have in all full 25,000 acres underlaid with this seam. The seam ranges from two to seven feet in thickness, is a "bright, glistening coal, having a columnar structure, and breaking in oblique lines, with partings of mineral charcoal, always tender, but frequently looking as if it had been subjected to severe strains." I have copied Prof. McCreath's description of coal B in Pennsylvania at various places, and every one acquainted with the Sewanee coal will immediately recognize the description. The variable thickness is also a feature of the seam in Pennsylvania.

These mines are located on the top of Cumberland Mountain, locally called Sewanee Mountain, and are connected to the Nashville & Chattanooga Railroad at Cowan by a railroad twenty-three miles long. The level of the plateau on which the houses of the company are built is 1,850 feet above the sea, thus giving pure and healthy air for the abode of the numerous operatives necessary around such an enterprise.

As previously stated, the company operating these mines is the Tennessee Coal, Iron & Railroad Company, of which J. C. Warner, Nashville, is President; N. Baxter, Vice-President; A. M. Shook, Tracy City, General Manager, and E. O. Nathurst, Tracy City, Superintendent of the mines at that place.

For the year ending December 31, 1882, the shipments amounted to 144,689 tons of coal, and 109,153 tons of coke. They have at this place 404 ovens. The average number of hands worked in 1882 was 600 convicts and 330 free labor. The average day's wages was: For convicts, $1.10; for free labor, $1.75. The value at the mines of the coal shipped was $205,000, and of the coke $180,000. The total amount of wages paid out in 1882 amounted to $180,000. New entries have been driven, old ones lengthened and much new iron rail laid down. The cost of mining has increased on account of the increase in length of entries. Free labor is paid 2 cents per pushel. Driving entries is paid for as follows: 75 cents per lineal yard for each foot of slate, and $1.00 additional for each yard of coal. Three hundred more hands are employed now than in 1880.

Fourteen entries from daylight into the ridge have been made; have in all about twenty miles of entries; size 8 feet wide, 5½ feet high. Full time is made all the year around, and sometimes work on Sundays. Public schools are taught at several points around the mines. The churches are Methodist, Episcopal and Roman Catholic. The cost of board is about $15 per month. Provisions are as cheap as in any part of Tennessee.

The main entry of the mine has been driven in on the seam one and a half miles; No. 2 has been driven in one and a quarter miles; No. 3 is in fifteen hundred yards, and No. 4 is twenty-two hundred feet long. These all belong to what is known as the old mine, where the seam was first opened. Two miles distant towards the eastern side of the plateau two other main openings have been made, known as the East Fork and Rattlesnake mines. The first of these is in about half a mile, the last about one mile. The entire length of entries in this mine from which it is available to draw coal exceeds fourteen miles. The largest average haul is one and three quarter miles.

The daily product for December averaged 1,200 tons of coal and coke, counting twenty-six working days. The convicts are tasked from five to seven boxes per day, a box holding sixteen bushels; hence the task ranges from eighty to one hundred and twelve bushels, depending on the thickness of the coal in which they are working. In three or four feet coal they easily dig their task in eight hours. Free labor is paid twenty-five cents per box when the coal is three feet and over, and thirty cents when under three feet, down to two feet. The men furnish their own oil and powder. No coal is mined under two feet. The seam sometimes swells to eight or ten feet in thickness, but seldom falls under two feet; the average of a large area is four feet. Miners easily make $1.75 to $2.00 per day. The Company keeps a store, but the men are paid in money. Some of them have accumulated property and bought little farms. Common laborers get $1.00 per day.

The convicts have good comfortable quarters and are well cared for. They are paid for all extra work. In 1882, over $500 per month were paid for extra work, to the convicts. Some of them accumulate money, comparatively a considerable amount. It is the interest of the company that the convicts should always be in good health and able to work, hence their physical condition is carefully guarded by keepers and wardens. They are permitted to sing as much as they please and have religious service every Sunday.

During the past winter this company has shipped large quantities of coal, but it intended to bring the business chiefly into the manufacture of coke. As previously stated, there are four hundred and four coke ovens, two hundred and four of these are at the East Fork mine, sixty-six at the Rattlesnake and one hundred and thirty-four at the old mine, and one hundred and thirty-four more are to be erected at the East Fork mine. The coke ovens vary in size, the old ones being ten feet in diameter and four and a half feet high inside, while the last built are eleven feet in diameter and eight feet high. The latter make the most compact and the heaviest coke. The first are usually charged with one hundred bushels of coal, the latter with one hundred and twenty-five bushels. The sixty-six ovens at the Rattlesnake mine are egg-shaped, nine and a half by fourteen feet, and and five and a half feet high inside. These are charged with eighty bushels of coal. In all the ovens the coal is allowed to burn forty-eight hours except that which remains over Sundays, this stays in seventy-two hours and makes a foundry coke. The average make is one hundred and five bushels of coke from one hundred of coal.

The opening, plant and equipment of the East Fork mine are among the best arranged of any in the State, and its history is a remarkable record of good management and rapid work. The entry was commenced January 2, 1880, and the ground was graded for two hundred and four ovens, the outside walls of the whole number built, one hundred of them fully finished and were burning coke by June 30th of the same year. At this place the coal as it comes from the mine is fed by elevators to the crushers, from thence it passes into large bins, of which there are three, each having a capacity of 4,000 bushels, and the fine coal dumped into buggies and charged into the ovens.

The ovens are of sandstone except the bottom, and cost $300 each; the bottoms are of red brick laid on their side. The coal is charged to a height of eighteen inches. The ovens are eleven feet in diameter, six to eight feet high, top opening thirteen inches in diameter, doors two and a half feet square. The pay for drawing ovens and loading on cars is forty cents per oven. Two men and two mules charge the ovens. The crushers have a capacity to crush and discharge into the bins every ten hours sufficient coal to supply three hundred ovens. The entire plant, engine, boilers, elevators and crushers is managed by two men, and cost only $7,500. The opening of this mine with one mile of main entry and T rail track all complete, cost $5,641.94. It is doubtful if a mine so well opened was ever developed at so small a cost.

At the old openings, a Stutz washer was erected. It was thought to have a capacity to wash six hundred to one thousand bushels per hour, and take out eight per cent of slate. The plant cost

$7,863.46, and there is great doubt if it possesses any advantage in treating this coal, as it is nearly free from sulphur. It is not now used.

The plant of this company for operative work consists of two crushing machines, four hundred and four coke ovens, eight hundred mine cars, eighteen coke-oven charging farries, 35,200 yards of small T rail track in the mines and many more of wooden tramway in side entries, also eight locomotives, and twenty-five miles of wide gauge track and two hundred cars. In addition are numerous houses, well fitted up machine and work-shop. The executive force consists usually of about nine hundred hands, and the company owns eighty-two mules. The general management of this company's business is under charge of Col. A. M. Shook, whose long experience in addition to natural ability gives him eminent fitness for the position, and the growth of the business to its present large proportions is ample evidence of the excellence of his management. Mr. E. O. Nathurst, the Superintendent of all the Tracy City operations, has been with the company for many years in the capacity of clerk, bookkeeper, cashier and now Superintendent, and no one can visit the premises without admiring the quiet, promptness and excellence with which every department is controlled.

SEQUATCHIE VALLEY.

The Sequatchie Valley region should not properly be a distinctive coal district, as all the coals so far mined in the mountains on its western side belong to the Sewanee series. In the geographical division of the State, this valley is placed in East Tennessee, but its mines, from their railroad connection, must be classed as belonging to Middle Tennessee. While the possibilities of coal mines in the western rim of this valley is great, the present development is very little, yet there is but little question that from this valley is presented the most favorable point for attack on the great body of the Sewanee seam, which exists in the counties of Bledsoe, Van Buren and Cumberland, from the fact that the upper ridges which contain that coal come nearer to the edge of the escarpment, in some instances being a mere receding bluff above the cliff. This seam of coal has been opened at many points from Victoria to the head of the valley above Pikeville. These openings are locally known as Deaken's, Stone's, Flagg's, Robertson's and other openings for testing, and home blacksmith's use have been made by W. H. Hart, Isaac Hopkins, R. S. Blackburn, J. Clark, and Hon. T. D. Northup. The seam of coal is stated to range from four to six feet in thickness.

In the mountain, on each side of the Little Sequatchie gorge, many openings have been made in the coal seams there existing, chiefly for local use. Most, if not all of them are below the Cliff, and hence of the Lower Measures. Among these is one opened by Hon. R. A. J. Ralston, said to be six feet thick, and the coal of excellent quality, the others are locally known as the Parmelly's bank, Prior's, Caldwell's Green's and Harris' banks. They are from six to twelve miles from the railroad, and though not now accessible, form an important part in the great storehouse of fuel to be drawn upon in the future, as though thus distant, the route to them is very favorable.

From Victoria to the northeast of the valley, the face of the mountain is comparatively regular, being broken only at Dunlap by the course of Brush Creek and thence to the head of the valley are only small streams, hence the disturbances of the strata occurring around the gorge of Battle Creek and Little Sequatchie are not likely to be met with in opening mines in that region. From Victoria to the head of the valley is about fifty-five miles, hence it is safe to assume that there is a regularity of strata for fifty miles. West of Dunlap the mountain is regular and unbroken, and contains the Middle or Sewanee coals for about twelve miles west, to beyond, and north of Tracy City. From the rim near Pikeville stretches out a vast area carrying the Sewanee seam far into Van Buren and north into Cumberland. The area of the Sewanee coalfield, more easily accessible to this valley than to any other point where transportation is available, cannot be less than four or five hundred square miles, which even at the former figure, and the usual estimate of 5,000 tons to the acre of merchantable coal from a four-foot seam would yield over one billion tons. The cheapness with which a railroad may be built up this valley, and the ease with which this vast amount of mineral fuel may be brought down to it certainly offers very attractive inducements to capitalists. In the valley is also a vast amount of the red fossil ore.

VICTORIA MINES.

These mines were originally opened by the Southern States Coal & Iron Company, an English corporation, but with all the other property of that company, were transferred in 1882 to the Tennessee Coal, Iron & Railroad Company. After a careful examination of all the vicinity of the mines, that company determined to abandon them, hence during 1882 but little has been done beyond drawing pillars.

In 1882 about 500,000 bushels (20,000 tons) of coal were mined, of this about 100,000 bushels (4,000 tons) were sold to the Fire Brick Works, and the rest was made into coke for the furnaces at South Pittsburg. There were 120 hands employed during the year, miners earning from $1.50 to $2.00 per day and laborers $1.00. The total amount of wages paid out was $31,000. The average price paid for mining was $2\frac{1}{4}$ cents per bushel; cost of living was somewhat higher than in 1880. John Frater is Superintendent, Post Office, Victoria.

These mines were opened in excellent style and at great expense, but have never been a success. They are not really in the main mountain, but in an arm between Little Sequatchie gorge and the Sequatchie Valley. The whole equipment, tracks, washing machinery, houses, etc., is of the most substantial character, and there is no better mine manager than Capt. Frater, but the most costly and perfect plant cannot overcome difficulties which nature has thrown in the way of cheap development, nor supply deficiencies which exist in her formations.

A seam of excellent coal, supposed to be the same, has been prospected about three miles farther up the valley, and is said to be of excellent quality, of good size, and appears to be as regular as is usual with the Sewanee seam. The seam as worked at Victoria contains an unusual amount of sulphur in nodules from the size of a pea to the size of a goose egg; these were extracted by washing. The thickness was also very variable, ranging from a mere thread to six or more feet; at one time more than 400 feet of hard sandstone was cut through. The seam dipped rapidly into the mountain, and there was a large amount of water to be pumped. The location of the mine was in every respect an unfortunate one, being just at the syncline of the mountain, and it is to be regretted that so costly and excellent an improvement has been abandoned.

LOWER COAL MEASURES.

In discussing the Walden's Ridge formations, allusion to the coal seams below the Seral conglomerate has been made. These seams, while having been worked at Daisy and Etna, yet have a minor importance in that region, but on the western brow of the coal-field they attain considerable thickness and have every appearance of being uniform, whether or not this be due to the fact that they have not above them in this region the great body of strata which covers them in the eastern part of the field, is a question for discussion; one fact exists, that the seams opened at Etna and Daisy and found at Soddy, under the main conglomerate, are there unreliable, while on the western face at Dade, Ga., one of these seams has been worked for more than ten years, and on the face of the mountain in Warren, White, Putnam, Overton and Fentress, they have been opened for neighborhood use, for that purpose worked for many years, and have proven to be reliable in thickness and of better quality than usual in the eastern field. These seams of coal have also been worked for many years on the Cumberland River, in Kentucky, and the coal was formerly boated down to Nashville, it there selling readily for two cents more per bushel than the West Kentucky coals. This coal is now mined on the Knoxville branch of the Louisville & Nashville Road, and is being brought to Nashville by Phillips, Randle & Co. Mr. Phillips states that the long railroad haul is against it, hence causing a higher price, but the superior quality makes it sought for. In that region these seams appear as they do in Tennessee on the western escarpment of the Cumberland Mountains, with but little more strata than the Seral conglomerate above them, and to the east gradually sinking to a lower level, so that the upper coals come in above; in the Jellico region to an extent of eight or more seams.

This peculiarity of unreliability on the southeastern face of the mountain is characteristic of these coals, also on the western side of Sequatchie Valley; the Battle Creek coal, mined there for many years, and highly esteemed for its good quality as a steam and grate fuel, cannot now be produced at a profit on account of the thinness of the seam. On the southeastern side of the valley, while coal has been opened there, yet, it is to be regretted, no depth of mining has been done, the reason assigned being that the pitch into the mountain was too great; neither has there been any opening in these lower seams at any point on the west side of the valley above Battle Creek.

The valley of Crow Creek runs at very near a right angle to the Sequatchie Valley, and, cutting the Cumberland Mountain nearly through, affords a means whereby the Nashville & Chattanooga Railroad crosses to the Middle Tennessee basin. In this valley, and far up on the mountain side, the strata is solely of the subcarboniferous limestone, but on the eastern side above, in a narrow strip of the umbral shales and sandstones, are several seams of coal. Two of these have been opened, and the coal shipped in small quantities to Nashville and other points. The quality is very good, the upper seam resembling the Old Etna, and like it, variable in thickness. A section made up the mountain by Dr. Safford, near Anderson Depot is as follows:

Sandstone, heavy, (the cliff)	120		feet.
1. Coal, from 2 to 5 feet, lustrous and laminated, with thin leaves of mineral charcoal, average	3		"
Fire clay	3		"
Shale	8		"
Sandstone	10		"
2. Coal and shale	10		"
Sandstone and sandy shale	10		"
Shale	1	to 6	"
3. Coal, no pyrite	2½	to 3	"
Shale with iron stones followed by debris, and the next rocks seen are the mountain limestone series.			

By comparison with the Etna section, before given, it will be seen that at that place there are five seams of coal, here only three. It corresponds with the Bon Air section on page 5, and 1 to 11 o the Sewanee section on page 11. A fine showing of this coal has been opened on a spur of the mountain called Frost Point, one and a half miles from Sherwood, which could be brought to the railroad by an incline and a branch track of not over half a mile. The coal is of the best quality for domestic use, would undoubtedly be a good gas coal, and mixed with Sewanee, have few superiors for generating steam. A large body of land in this region is under the control of Col. N. E. Alloway, of Nashville. Why Nashville is not supplied with coal from this region seems unexplainable. The coal is of the best quality, far superior to that from Hopkins county, Ky., so largely used, and the freight distance is but little greater, the distance from Sherwood to Nashville being 97 miles, from Anderson 102 miles, while from Nashville to the nearest mine in Hopkins county, Ky., is 85 miles. From Sherwood to Chattanooga is 54 miles, and from Anderson 49 miles.

Further north, beyond the railroad tunnel, these coal seams are found in similar order, and have been worked to considerable extent. The largest of these operations is called the University mine, being very near or in the bounds of the lands belonging to the University of the South. It is conducted by Mr. Charles Richardson, and the coal is of excellent quality and much sought after for domestic use. He informed me that he had already engaged forty car-loads to be sent to Fayettville for next winter's use. Col. C. D. Jones also owns a mine of the same coal only 1,700 feet from the railroad, further northeast from the University mine; still further on about three miles, and one and a half miles from the railroad Mr. W. B. Gibson has opened a seam which is no doubt the Jackson seam below the Sewanee. Mr. Richardson's mine is 2½ miles from the railroad.

A few miles beyond the Tracy City mines a long arm of the mountain juts out far into the valley. It is the water divide, and is a distinct ridge far into the plateau country in which Manchester and Tullahoma are located. It towers high in the air, and its summit is capped with the remnants of the Lower Coal Measures, but farther in towards the main mountain where the coal seams are found they are thin. Following the mountain, however, in its northeastern course, in the bluffs near the head of Collins River, the Lower Measures thicken and the three seams of coal appear of workable size, while the Sewanee seam is also found in the strata above the cliff and very near the escarpment. It has here been worked in a small way by Lawson Hill, Esq., and the coal hauled to McMinnville. Still farther northeast, up a fork of Collins River, the several seams have been opened by Mr. Barnes, and at Myers' by Houchin & Biles, and at McCorckle's by Womack & Doty, and the coal hauled to McMinnville. The same characteristics found elsewhere exist here, the sub-conglomerate or Lower Measure coals are firm and cubical, bearing transportation very well; the Sewanee seam has its crushed appearance, and is highly esteemed as a blacksmith's coal. From this point to beyond Sparta, as far as my actual observations extend, these Lower coals are of very regular thickness, but southeast of Caney Fork the Sewanee seam recedes from near the mountain brow and probably exists over a small area on the east side of that stream.

In the neighborhood of Sparta a number of openings have been made in the Lower Coals, from some of them the coal has been mined for many years. Bon Air is an arm of the mountain, jutting out in the valley something like Ben Lommond, but not so far; it is capped with the Seral conglomerate, the true cliff rock, and below it are found the Lower Coals. Near the point they appear to be thin, but passing east, on either side of the arm, they become thicker, and have been opened on both sides. A large body of this land formerly belonged to Hon. G. C. Dibrell, but is now owned by an association of gentlemen known as the Bon Air Coal Company. They own about six miles on the face of the mountain; two old openings, known as the Fitzwater Bank and the Trowbridge Bank, are both on their property; Little's Bank is on a small tract within their boundary. At the Fitzwater Bank two seams have been opened, and a third is known to exist immediately under the conglomerate. The last is undoubtedly the same as the old Etna, as seen by me at various points its characteristics are identically the same. At Little's Bank two seams of coal have been opened, the upper 3½ feet and the lower 44 to 48 inches in thickness. At Fitzwater both seams have been opened in past times, but only one is now worked; it is 4 to 4¼ feet thick, and is a hard, firm coal of excellent quality. Still farther east, near the head of Blue Spring Cove, is the Trowbridge Bank, where the coal has been opened 3 to 3½ feet thick. This is also on the Bon Air Company's property. On the opposite side of the cove the coal has been opened by Mr. King 2¾ to 3 feet thick.

Following on around to the northeast is an opening known as Officer's Bank. This property is now owned by Kinsey & Butler, who bought the land for the timber. The seam opened is the lower of the three, usually found in the Lower Measures, and is from 3½ to 4 feet thick. I found the second or middle seam, but no opening beyond a little of the outcrop was made; the upper or cliff seam shows plainly under the cliff rock only a few feet from the main road. These gentlemen have full five miles of the face of the mountain, and, while only purchased for timber, they have also obtained a valuable coal property. The Calf-killer River rises far up in the mountains of Putnam County, and so divides the coal-bearing measures as to give access to an immense amount of that mineral.

It remains a question for the future to solve whether these coals can be brought to the Nashville market. The distance from Nashville to Sparta is 137 miles, from thence to the Fitzwater Bank a route has been surveyed, and the coal can there be reached in five and a half miles. The superior quality of these coals for domestic use, for the manufacture of gas and for steam generation, is undoubted, and it is plain that if they can be brought into the Nashville market they will become a dangerous rival to the Kentucky and Alabama coals, now almost entirely used there. Back from the brow of the mountain, a greater or less distance all the way from the head of Collins' River to Caney Fork, the Sewanee seam appears in the superimposed ridges, many of them extending over a large area. In some it has been opened and shows a thickness ranging from $3\frac{1}{2}$ to 6 feet; one of these at Scarborough's mine is somewhat noted, coal having been taken therefrom for many years. The long spurs extending out from the mountain and the gorges, cut down by the various forks of Caney and Rocky rivers, give routes for railroad lines to reach the top of the rim and to penetrate the region of this excellent coking coal. In writing of Sequatchie Valley, the great area underlaid with it has been spoken of, and the same will apply to this side of the mountain. The area possible to be reached from either side is capable of yielding hundreds of millions of tons, and it will simply be a point for the railroad builders to decide from which is the cheapest and most advantageous means of reaching the coal and bringing it to a main line. It is plain, however, that it appears more to the interest of the Nashville & Chattanooga Road to have the connections made to their McMinnville branch, over which traffic is sparse, rather than to the Jasper line, a part of which at least is heavily burdened. The fact is, however, that the good coking coal exists in large quantities, and it will soon be an absolute need of the public that access shall be had to it.

Northeast of Sparta, in the counties of Fentress, Putnam and Overton, the same seams of Lower Measure coals are found, and on into Kentucky. A fine development of these seams is shown in the bluffs along the South Fork, as also on Obed's River. Mr. Cyrus W. Clark has been for some time boring on a large body of land, owned by himself and others in Fentress County, and has found a very persistent seam of coal 4 to 4½ feet thick. Formerly, coal was shipped out of Obed's River to Nashville in flat-boats. The great possibility of the development of all this region exists in the construction of a trunk line east from Nashville, either to the Cincinnati Southern alone or to it and beyond to Knoxville.

It is possible that the Sewanee seam may be found in the ridges in the southern part of Fentress and Overton counties, but the great elevation of the main conglomerate and the demolition consequent thereon, has swept it away from the middle and northern section. As previously stated, nearly the entire area of Cumberland County is underlaid with this seam of coal, which has a general rise to the north.

It is to be regretted that some one of the many wealthy owners of large tracts of land in the plateau country of Cumberland, Van Buren and Bledsoe, have not thought fit to bore down through the strata and determine whether the Lower Measure coals underlie all that region. If it be a fact that these coals do thus exist, and in the thickness which they appear all around the rim, from Ben Lommond to Kentucky, our store of mineral fuel becomes incalculable. The fact that they are found in Walden's Ridge gives good ground to base the theory that they do thus exist under this vast area.

APPENDIX.

IRON ORES OF TENNESSEE.

The State of Tennessee contains every variety of iron ore known to commercial use, except the Spathic Carbonate. The boundary of the Magnetic ores, and of the azoic Hematites, is not extensive, yet in the limited area where found, the magnetic ore exists in large quantity. The mass of unaltered deposit ores, however, is beyond the possibility of any accurate computation, and the area in which they are contained comprises nearly three-fourths of the State.

Geographically, these ores may be classed as the East Tennessee Iron Region, the Cumberland Mountain Iron Region and the Middle Tennessee Iron Region. Geologically, they belong to the Metamorphic, the Lower and Upper Silurean, the Sub-Carboniferous and the Carboniferous periods. Physically, they are vein, stratified and deposit ores; and in practical nomenclature of ores, they are magnetic, specular, red hematite or really hematite, limonite, frequently called brown hematite, red fossil or lenticular red hematite, and carbonate of iron. Of these ores, those now used in the State are only the limonites and red fossil. The magnetics have been mined, and some years ago used, in forges, but none have yet been used in blast furnaces in this State, except as an experiment, though large quantities have been shipped to Allentown, Pa., and there used with good results. The azoic hematites are known only by small openings and specimens of more or less size. These two ores have been found only in the counties of Johnson, Carter, Unicoi and Cocke, but are thought also to exist in Sevier, Blount, Monroe and Polk.

The limonites are found over the largest territory, and have been most generally used of the two chief ores of iron. They are found in nearly every county of the State in greater or less quantities, from the North Carolina line to the sand belt which borders on the Mississippi River. In some counties the quantity is enormous, in others only scattered specimens, and the quality is equally variable; some beds are almost chemically free from phosphorus or sulphur, while in others those injurious elements are found to a greater or less extent.

In East Tennessee this ore lies in a series of ridges running northeast and southwest; its greatest development being on the east side, on the western slopes of the Chilhowee and Unaka Mountains and their tributary ridges. Throughout the entire breadth of the State, in the counties of Johnson, Carter, Unicoi, Washington, Greene, Cocke, Sevier, Blount, Monroe and Polk, there may truly be said to be one continuous bed of limonite, at some points in immense masses like stratified or boulder rocks; at others intermingled with the soil, but yielding large quantities of ore when subjected to the process of washing. The ores of this lead are all in the lower silurean, and usually lie in slates, or between the Chilhowee sandstones and the dolomites of the Knox or Quebec periods, frequently intermingled or deposited between masses of the latter. In this position, it is found in a matrix of red or yellow clay, from the size of coarse sand to large boulders. These are the ores from which a large part of the iron of the United States was made in times past, and many beds are now worked in Pennsylvania, New York and Massachusetts, from which ore was taken a hundred years ago. The unsystematic and robbery-like character of obtaining the ore from many of the banks in Tennessee has greatly impaired their value, and in some cases apparently exhausted the supply of ore.

The limonite of this lead varies very greatly in quality, some being very free from any impurity, almost pure hydrated oxide of iron, but the greater part contains silica, alumina, phosphorus and sulphur in greater or less proportions, none to such an extent as to make it worthless. In some beds manganese prevails in such proportion as to make the manufacture of speigeleisen or ferro-manganese a possible source of profit. These deposits become more vast in size toward the southeast corner of the State, and the deposits on Tellico River and Gee's Creek, between the Little Tennessee, Hiwassee and Ocoee rivers, challenge the admiration of the geologist and practical iron manufacturer.

At intervals in every ridge of the Knox dolomite formation, beds of limonite are found. Some of them appear to be of considerable extent, though but few of them have been opened. When opened, the quality of the ore has proven to be good. On the summit of the Walden's Ridge, at various points from Emory Gap to Careyville, beds of limonite are found, which are no doubt the result of local change of the carbonate of iron of the coal formation.

The largest body of limonites in the State is found in Middle Tennessee, in what has been usually called the Western Iron Belt. This vast deposit covers irregularly an area forty miles wide and extending entirely across the State from North to South. It comprises the entire area of the counties of Wayne, Lawrence, Lewis, Perry, Hickman, Humphreys, Dickson, Houston, Montgomery, Stewart, and part of Benton, Decatur and Hardin.

The surface geology of this region belongs to the sub-carboniferous. It is in fact the counterpart of the Cumberland plateau of the east, with the coal measure rocks and the upper limestones swept away. The general elevation of the corresponding strata underlying the coal measure rocks is but a few feet more than that of Lawrence and Hickman counties. Almost at an identical level, on each side of the Middle Tennessee basin, occur the same characteristic rocks. The vast body of coal which once may have extended from Kentucky to Alabama is gone, but deposited in its underlying strata; from the slow action of ages, now remain immense bodies of iron ore, in quantity and quality hardly surpassed by any like area in the United States. In phosphorus and sulphur these ores frequently go down to a mere trace, while they never rise to such an extent as to be in the slightest degree injurious for the very best grades of foundry irons.

The location of this ore has been stated to be an elevated plateau-land, yet it is well watered with many springs, and is also intersected with streams which flow west from the Middle Tennessee basin, being cut through on the north by the Cumberland River, while the western edge is intersected from north to south, the entire middle of the State, by the Tennessee River. All these streams cut down through the sub-carboniferous strata into the lower limestones, thus affording ample facility for obtaining flux in the manufacture of iron. The two great rivers named also afford cheap transportation to markets, while other means of transportation and access to this region is afforded by the Memphis branch of the Louisville & Nashville Railroad through Motgomery Stewart, and Houston counties, the Nashville and Northwestern through Dickson, Humphreys and Benton, a narrow-gauge south from Dickson

Station into Hickman County, and the railroad from Columbia through Lawrence County to Florence, Ala.

This ore has been almost entirely used for the manufacture of iron with charcoal, and there are now six furnaces operating in this region. All use charcoal for fuel: three are cold blast and three are hot blast. Notwithstanding its contiguity to reliable and cheap transportation, but little of this ore has ever been shipped to market in other States, nor to any coke furnace in this State. The connection by the Duck River Valley Road from Columbia to the Nashville, Chattanooga & St. Louis Railroad, already completed, gives an outlet for this ore directly to the coal, and it will probably, at some future day, be shipped to furnaces on the line of that railroad or in Chattanooga.

It has been thought that of all this region, the county of Hickman excelled in the quantity of ore it contained, but later investigations show that Wayne and Lawrence have nearly as great bodies of ore, and vast beds have been found in Houston, Dickson and Humphries, which, if known at all, had only a local reputation. Many beds abandoned years ago have also been reopened and proved to have been by no means exhausted. Some of these beds have been worked to a depth of seventy feet below the original ore surface, and the bottom of the ore bearing strata not yet reached. In some of the beds that have been worked the amount of debris to a ton of ore amounts to three to one, in others it is much less, and frequently the ore is met with in enormous masses of many tons in weight. The area covered by the formation in which this ore exists is over 1,200 square miles, and it is fair to state that not less than one-twentieth is underlaid with iron ore, for there is not an east and west ridge but has one or more beds in its length, while there is not a north and south ridge which is not almost throughout its length a depository of ore, usually in large bodies.

The amount of ore in the counties comprising this belt is simply incalculable; one geologist estimates the supply in Lawrence county alone, on and near the line of the Nashville & Florence Railroad at one hundred million tons. The county of Wayne being out of the general line of travel and present transportation, has not received the attention given to some of the other counties, but it contains many large and valuable beds of iron ore. The beds of

Lawrence are famous, the most noted are the Tucker, Herren, Wright, Nixon, Powell, Smith and the Napier Furnace banks. The Nixon bank, in Maury, near the Lewis county line, is a wonderful deposit of ore, in Lewis are many good ore banks undeveloped; in Perry, on the east side of Buffalo, are vast bodies of ore lying quiet as nature placed them, while on the west side, in addition to the vast bodies of undeveloped ore extending in a north and south ridge through the whole county, are the banks belonging to the old Bradley or Cedar Creek Furnace property, great in extent and of the best quality. In Hickman, south of Duck River, the Etna banks are famous for their extent and the good quality of the ore; the Hickman County Company's beds, near by, are equally as good. There are many others undeveloped. North of the river are numerous beds, those developed belonging principally to the Warner Iron Company. In Humphreys there are many good beds undeveloped, while in Dixon are a number of old works from which thousands of tons have been taken and much more still to be found, while there are also numerous undeveloped banks with the appearance of vast quantity of ore. The same remarks are true of Houston, Stewart and Montgomery. In the latter county is found ore of the few large and apparently inexhaustable beds of the variety called pipe ore.

This region is the true home of the charcoal iron-maker. Firstly, the ores are abundant, easily and cheaply mined, and are easy to smelt, limestone is convenient, there are hundreds of thousands of acres of timbered land which will never be of any value for farming purposes. Experience has demonstrated the fact that the timber cut off this land will be renewed by nature of ample size for recutting every thirty years. The old coalings of Cumberland, La Grange, Etna and of the many abandoned furnaces prove this. This land is never likely to become so valuable that it will not pay to let it grow up in timber, while at the same time it is growing up in timber it affords excellent pasturage for cattle and sheep. By the use of cheap narrow gauge roads the wood can be brought to the Furnace from considerable distance at a very low cost. The peculiar trend of the streams of this region and the location of the ore in the hills between them, give a judiciously selected site vast advantages, as not only may the wood for long distances be brought down hill, but also the ore in the same way and on the same road.

Hence in this region it is possible to reach a minimum cost for the raw material.

An examination of the map shows the peculiarity alluded to. Shoal Creek, in Lawrence county, receives all its tributaries from the north; Buffalo River, running to the west, receives its affluents from the south, but turning its course north receives them from the east; Duck River is the same, and so to a great degree is the Cumberland. Between any of the tributaries to these main streams are vast bodies of ore and immense areas of virgin timber.

Near Brownsport, in the County of Decatur, occurs a bed of limonite, probably extending over a very considerable area, which is not referable to any of the the formations in which that ore has elsewhere been found in Tennessee. The ore occurs stratified in layers and masses just beneath the black shale of the Hamilton period, Devonian age, and rests immediately on the Helderberg limestone. Immediately above the black shale is the siliceous group of the sub-carboniferous. A furnace was once operated at this locality, and the stack and some of the houses are still in good order; the machinery is excellent. The ore at that point is in large quantity, and it appears to exist in the same geological position at about the same elevation over a considerable section of the surrounding country. The furnace, though thus eligibly located, was badly managed and has been idle for many years, being tied up in the meshes of the law. The nearness of this site and ore to the cheap transportation afforded by the Tennessee river should cause it to be utilized. The ore undoubtedly exists in great quantity over a large area of country—up and down the river.

Along the western foot of the Cumberland Mountains on the eastern side of the Middle Tennessee basin, in a formation identical with that where the ores of Stewart, Montgomery and Hickman are found, exist some beds of limonite, the extent of which have not been fully determined. They are found chiefly in the counties of White, Warren, Putnam and Overton. At several points these beds appear to be of valuable extent, but no exploration has been made sufficient to test the quantity. The McMinnville branch of the Nashville & Chattanooga Railroad, now in course of extension to Sparta, will afford means of transportation and access to this region.

(101)

RED FOSSIL ORE.

The next ore to be considered, and though occupying a less area, probably not less extensive in quantity, belongs to the true hematite series, and is known to mineralogy and the manufacturer as the red fossil ore, but is known locally in Tennessee as dyestone. It is almost entirely confined to East Tennessee, but almost three-fourths of the pig-iron made in the State since 1870 was made from it.

The geological position of this ore is in the Clinton group of the Niagara period, below the black shale of the Devonian age. But in East Tennessee, all along the western base of the Cumberland Mountains, from Chattanooga to Cumberland Gap, the two strata are found in close conjunction, and where one exists it is there certain that the other is to be found in that vicinity, though it may be covered with drift. This ore is one of the most persistent strata of the Appalachian geological system. It is found in New York, bordering Lake Ontario, curving northward on the west and southward on the east, sinking there beneath the Hamilton shales and slates, rising again in Pennsylvania, and continuing thence in an almost unbroken outcrop southwest into the heart of the State of Alabama. The seams of ore in this State, however, are much thicker than in Pennsylvania; and besides the regular continuous seam at the foot of the Cumberland Mountain, there is another seam almost as continuous, and at places much thicker, in what is called White Oak Mountain, a high ridge entering the State from Georgia, in the county of James, and passing northward into Virginia, though the northern end, in the county of Hancock, is called Powell's Mountain. This is the Montour ridge of Pennsylvania. This ridge in Pennsylvania is only twenty-seven miles long, and from it in 1864, Prof. Rogers states that twenty furnaces, making sixty thousand tons of iron per annum were deriving their supply of ore, and in 1881 there were still nine large furnaces deriving their supply in whole or part from this same ridge. The White Oak Mountain has a continuous length in East Tennessee of over one hundred miles.

This red fossil ore is also found in several detached ridges, from three to ten miles long, which lie parallel with the White Oak Mountain, at intervals, in a general southwest and northeast direction.

This ore is less variable in quality than the limonites, and the analysis of a specimen from one point in a leading ridge will usually be identical with that from another point ten, twenty or fifty miles distant. Below water level, the ore on the White Oak Mountain, and at a certain depth the ore in the seam at the foot of the Cumberland Mountain, becomes poorer in iron and richer in lime. Hence, for the present, mining is stopped when this hard and poor ore is reached; the proper course would be to mix it, as done in Pennsylvania, with the richer soft ore from near the surface.

Two other bodies of this ore are of great extent in East Tennessee, but detached from the East Tennessee Valley proper. These are in Elk Fork Valley and Sequatchie Valley The former is about twenty-five miles long and extends into Kentucky; the latter is about sixty miles long and extends into Alabama. Throughout the whole length of these valleys the red fossil ore appears, dipping slightly to the east. On the opposite side of the mountain, at its eastern base, along the foot of Walden's Ridge, the ore dips to the west, hence if the ore is continuous for the eight to ten miles of distance under the intervening carboniferous strata, the amount of iron ore thus stored away for future use is simply enormous. The ore on the east side of the mountain is three feet thick and in the detached seams much thicker. Therefore, even if containing only 30 per cent of iron, the amount of available ore the seam would yield, to capital invested in scientific mining, will equal if not surpass that of any known deposit of iron ore in the world.

At present the mode of mining this ore is to get it on the cheapest plan possible, without the slightest reference to the future. In the seam at the foot of the mountain it occurs in a series of knobs, with short, narrow valleys between them. The ore is robbed from the knobs by rough tunnels as long as they think it pays, and then that knob is abandoned and another attacked. No mining is done below the level of the little branches. In White Oak Mountain the dirt and shale is stripped with picks and shovels off the seam of ore until the wall of shale reaches a height or thickness or six or eight feet; the stripped ore is then taken out and the rest abandoned. In so-called worked out leases near Ooltawah, are thousands of tons of ore which, by intelligent mining can now be gotten out as cheaply as has been any which had the thinner covering. The price of this ore in Chattanooga is $2 to $2.50 per ton.

On the mountain seam are now located three furnaces, two at Rockwood and one at Oakdale. One furnace at Chattanooga derives its supply from the White Oak Mountain near Ooltawah, and from the mines up the river, and South Pittsburg and Cowan furnaces get their ore from Ooltawah and up the river at Half Moon Island, and A. Welcker's, and some from Alabama.

The seams of this ore have very superior facilities for transportation. The Tennessee river runs parallel between the White Oak Mountain seam and that of Shin Bone Ridge, at the foot of the Cumberland Mountain. The latter has also the Cincinnati Southern Railway in a few hundred yards of it for nearly seventy miles. It is also accessible by the Knoxville & Ohio Road at Coal Creek and Careyville. The White Oak Mountain ore is cut through by the East Tennessee, Virginia & Georgia Railroad near Ooltewah and also by its Red Clay extension, and by the Knoxville & Ohio branch of that road from Knoxville to Kentucky near the town of Clinton. The Tennessee river also cuts through the Half Moon Island bed for a distance of ten miles. A system of cheap narrow-guage roads would bring to the river and railroads in short distances a large amount of ore now too far distant for hauling by teams. The red fossil ore has not been found in any part of the Middle Tennessee region. In Overton county a hematite ore is found, locally called dyestone, but it is not the same as the East Tennessee dyestone, nor is it known to exist in large quantities. In the county of Wayne are three knobs which contain a large amount of red iron ore, not properly a hematite. Its geological position has not been exactly determined. The location is near Clifton, on the Tennessee River, and the ore is of good quality. It was once used in a furnace located near by, and made good pig iron. Some of it has been shipped off and made into paint.

CARBONATE OF IRON.

The third most important ore, as respects quantity, in the State of Tennessee, is the carbonate of iron of the Coal Measures. This is in England and Europe one of the chief ores from which iron is made. It is used to some extent in Ohio and Pennsylvania, but as yet not at all in Tennessee, though it is one of the most abundant and easily worked ores. There are points in the Tennessee coal field where it can be mined very cheaply. It is found in the State underlying the coal seam, worked at Coal Creek and at Careyville; at the latter place it is specially abundant. There are a number of layers of it in the Tennessee coal-field.

Large quantities of coal are also found in the Umbral or Lower Coal Measures of Marion, Grundy, Franklin, Warren, White, Putnam and Overton counties.

MAGNETIC ORES.

The least abundant, but most valuable iron ores of the State, are the ores found in the metamorphic rocks, from which Bessemer steel pig may be made. These are the hematite and the magnetic. They are found at intervals in the strata just edging on the Potsdam sandstone and in the hornblendic gneiss of Carter and Johnson counties. The hematite has not been developed to any special extent; hence its quantity is not known. In Sullivan and Carter counties, in the foot-hills of the Holston Mountains, is found hematite ore of very compact structure. It has been used in forges and in small charcoal furnaces and made good iron, but no sufficient exploration has ever been made to test its quantity, though small pieces of it are scattered over a large area of country.

The magnetic ore exists in a limited area, but is in large quantity and of excellent quality. Little beyond explorations for the investment of capital, and a little digging for forges has been done in this State, but beyond the North Carolina line, very extensive excavations have been made by the owners of the East Tennessee, & Western North Carolina Railroad, and an immense amount of ore uncovered. That railroad is now completed from Johnson City to those mines in North Carolina, and must also eventually be the means of developing the ore of Carter county. In the eastern part of Johnson county magnetic ore is also found, but transportation is so far distant that there is no likelihood of its development for many years.

The following are the iron furnaces of Tennessee using coke for fuel:

Oakdale Iron Company, Jenks P. O., Roane county, Tenn.: Hon. John G. Scott, President, Jenks P. O., (one stack).

Roane Iron Company, Rockwood P. O., Roane county, Tenn. (two stacks), H. S. Chamberlain, President, Chattanooga, Tenn.

Chattanooga Iron Company, Chattanooga, Tenn., (one stack). Wells, Manager, Chattanooga, Tenn.

Tennessee Coal, Iron and Railroad Company, South Pittsburg, Marion county (two stacks), J. C. Warner, President, Nashville, Tenn.

Tennessee Coal, Iron and Railroad Company, Cowan, Tenn. (one stack), J. C. Warner, President, Nashville, Tenn.

The combined product of these furnaces is about 400 tons of pig iron per day.

Citico Furnace Company, E. Doud, General Manager, Chattanooga, Tenn., are in course of erection; a stack built with the latest improvements, which will make one hundred tons per day.

The furnaces now in operation in the State of Tennessee using charcoal for fuel are:

Napier Furnace Company, Chief, post-office, Lawrence county, Tenn.: J. E. R. Carpenter, President, Columbia, Tenn., makes coldblast car-wheel irons. (Out of blast.)

Warner Furnace, Warner, Hickman county, Tenn.: J. C. Warner, President, Nashville, Tenn. Makes hot and cold-blast carwheel iron.

Drouillard Iron Company, Cumberland Furnace, Dickson County; J. P. Drouillard, President, Nashville, Tenn.; hot-blast charcoal iron.

Cumberland Iron Works Company, Bear Spring Furnace, Stewart County; J. P. Drouillard, President, Nashville, Tenn. Makes coal-blast charcoal iron for car wheels.

La Grange and Clark furnaces, La Grange Iron Company, Postoffice Danville, Houston county, Tenn.

The combined product of these furnaces is about one hundred and five tons per day.

EXPLANATORY.—In the original plan of this volume, the intention was to include in it a sketch of the Western Iron Belt, information derived from various explorations, made under direction of Dr. A. W. Hawkins, then Commissioner of Agriculture, Statistics and Mines, but to do so, and at the same time make the volume so complete on the coal of the State as to meet the public demand, would exceed the sum left by Commissioner Hawkins to pay for its publication; hence a mere resume of the iron ores of the State is given here, and the data of those explorations made in the service of the State will be given to the public which is entitled to them, in the future, in some other form.

HENRY E. COLTON.

MANUFACTURE OF COKE.

The following is an extract from a paper on Coke by John Fulton, General Mining Engineer of the Cambria Iron Company, published in Vol. 4 of the Pennsylvania Geological Survey:

In all coking operations the work to be accomplished is to *expel the gaseous elements of the coal*, retaining the carbon and ash which constitutes the *coke*.

It is thus evident that the quantity of coke obtained from any coal cannot exceed the sum of its carbon and ash.

On the other side it is rarely found that coke can be made without the loss of several units of carbon, depending on the quality of the coal and the method of coking it.

The minimum loss of carbon should be made in coals having a large volume of hydrogenous matter; in other words, holding a sufficient amount of gaseous product to supply the necessary heat for the operation of coking without using any of the carbon.

The maximum loss of carbon would result in coking a dry coal, or one holding a small percentage of gaseous matter, thus requiring the burning of carbon to supply the necessary heat.

These considerations lead in the outset to an inquiry into the requisite qualities in a good coal for coking.

It might be expected that all bituminous and semi-bituminous coals would produce good coke. That such is not the fact is now becoming clear to those interested in this industry. The difficulty hitherto in getting light on the requirements of a good coking coal, and the principles of coking it, consisted in the loose statements of the advocates of the several kinds of ovens, who seem determined to make *them* the prime element in governing the quantity and quality of the coke produced. The quality of the coal used, or contemplated to be used, being regarded as an unimportant factor in this consideration.

There is, doubtless, great economy in the use of proper ovens in coking coal, but under all this *the character of the coal* is the *prime factor in determining the quantity, quality and structure of the coke.* And this is true whether the coal is coked in the most improved oven or in the primitive open air "pits" or mounds.

The value of ovens is confined rigidly to the economy of labor in the process of coking, and in the saving of carbon.

Blast furnaces demand the fuel to be pure, compact, tenacious, of uniform quality, and as free from moisture as possible.

It is evident that the calorific power of coke is derived from its carbon, and hence the purest coke will produce the greatest heat. This requirement of pure dry coke is the more evident when it is considered that all foreign matter and moisture not only do not contribute heat but require the expenditure of it, in disposing of the extraneous matter in the slag and vaporizing the moisture.

It is manifest that as *the character of the coke* is determined *by the quality of the coal used,* the latter should receive very careful examination before expending largely in plant for coking.

The first requirement in the production of good coke is a pure semi-bituminous coal—coal having small quantities of slate, sulphur or phosphorus.

The second requirement is, that it contains a sufficient proportion of volatile or gaseous matter to supply the necessary heat in coking, without the expenditure of carbon.

And thirdly, that the coal produces a coke of sufficient tenacity to sustain, without crumbling, the burden and blast of the furnace, and to inherit *an open cellular structure, to facilitae its impregnation and solution by the carbonic acid gas in the furnace.*

Three belts of semi-bituminous coals have been thoroughly tested in the production of coke for blast furnaces—the Connellsville coke region in the west of the State, Bennington, on the crest of the Alleghany mountain, and the outlying coal field of Broad Top in the east.

These furnish the three types of the best qualities of coking coals of the State. The analyses of thes typical coals are given as a means of comparison and siandards of qualities in coking coals.

THREE TYPES OF THE BEST QUALITIES OF COKING COALS OF THE STATE.

	CONNELLSVILLE. Pittsburg seam.	BENNINGTON. Miller.	BROAD TOP.	
			Barnet.	Kelly.
	a.	b.	b.	c.
Fixed carbon	59.62	68.50	74.65	71.12
Ash	8.23	8.00	7.50	7.50
Volatile matter	31.36	22.38	16.00	19.68
Sulphur	.784	1.12	1.85	1.70
	100.	100.00	100.00	100.00
Coke	68 per cent.	76 per cent.	81 per ct.	78 per ct

ANALYSES OF OTHER COALS FOR COMPARISON

	PITTS'BG SEAM. Irwin's mines.	ENGLISH COAL. Durham.	WELCH COAL.
	d.	e.	f.
Fixed carbon	61.45	83.27	80.50
Ash	5.80	1.52	6.50
Volatile matter	31.71	8.21	12.10
Sulphur	1.04	Not given.	0.90
	100.00	100.00	100.00
Coke	66 per cent.	84 per cent.	86 per cent.

DRY OR NON-CAKING COALS.

	Mahoning Vall'y Pa.	Brazil, Indiana.	Straitsville, Ohio.
	g.	h.	i.
Fixed carbon	64.30	57.20	55.60
Ash	1.95	1.90	6.94
Volatile matter	32.73	40.15	36.50
Sulphur	1.02	0.75	0.86
	100.00	100.00	100.00
Coke	65 per cent.	58 per cent.	61 per cent.

The property of caking or not caking in the soft coals does not appear to be clearly made out yet. It does not depend on the amount of volatile matter, for the non-caking coals possess this in the largest volume. Nor, as a general rule, does it appear that the coking property increases with the increase of the hydrogen and oxygen, but rather on the presence of different kinds of bitumen, or to the chemical constitution of the coal as respects the manner in which the gases are combined with the carbon, This combination producing hydro-carbon will account in part for the loss of carbon in coking, but not all.

Ordinary analyses fail to indicate the essential qualities of a good coking coal. They are highly useful, however, in exhibiting the carbon, ash and sulphur, thus clearly indicating the strength and purity of the coal.

The only sure method in the determination of the adaptability of coal for coking, is to have a quantity of it made into coke, and a study of its physical and chemical properties carefully made.

Other conditions being satisfactory, coal can now be *cleansed from an excess of slate and sulphur by a process of crushing and washing.*

At the Cumbria Iron and Steel Works at Johnstown, Bradford's cylinder breaker with the usual jigs are used.

By this washing operation, many coals can be made into good coke, which otherwise would not prove satisfactory.

With coals adapted to coking, the result can never be doubtful, whether the coke is made in open heaps, Beehive ovens or Belgian ovens.

The primitive mode of coking coal in heaps or mounds, very naturally grew out of the method employed in making charcoal in conical mounds for furnace use.

The plans are essentially the same; but in the case of the coal it has been gradually improved, mainly in respect to uniformity of quality of coke and saving of carbon.

The coke yard is prepared by leveling a piece of ground and surfacing it with coal dust. The coal to be coked is then arranged in heaps or pits, with longitudinal, traverse and vertical flues; sufficient wood being distributed in these to ignite the whole mass.

Beginning on a base of 14 feet wide, coal is spread to a depth of 18 inches, A. On this base the flues are arranged and constructed as shown in the plan—the coal being piled up, as shown in section B. The flues are made of refuse coke and lump coal, and are covered with billets of wood. When the heap is ready for coking, fire is applied at the base of the vertical flues, C, C, igniting the kindling wood at each alternate flue.

As the process advances, the fire extends in every direction, until the whole mass is ablaze. Considerable attention is required in managing this mode of coking, in diffusing the fire evenly through the mass, in preventing the waste of coke by too much air at any place, and in banking up the heaps with fine dust as the operation progresses from base to top.

When the burning of the gaseous matter has ceased, the heap is carefully closed with dust or duff, and nearly smothered out in this way. The final operation is the application of a small quantity of water, down the vertical flues, which is quickly converted into steam permeating the whole mass. This gives coke with the least percentage of moisture, if carefully applied.

The time necessary for coking a heap with the Bennington coal, is from 5 to 8 days—depending mainly on the state of the weather.

It will be shown that coke made in this way is beyond any doubt excellent.

The yield of coke, accurately determined at Bennington and Hollidaysburg, is as follows:

BENNINGTON.

Coal used,	56.87 gross tons.
Coke drawn,	33.63 " "
Loss,	23.24 gross tons.

Yield of coke, 59.1 per cent; loss, 40.9 per cent; 1.67 tons of coal to 1 ton of coke.

HOLLIDAYSBURG.

Coal used,	63.80 gross tons.
Coke drawn,	38.02 " "
Loss,	25.78 gross tons.

Yield of coke, 59.00 per cent; loss, 41.00 per cent.

The yield at both places is substantially the same, 59.00 per cent, exhibiting a loss of 24.00 per cent of the carbon contained in the coal.

The *Beehive oven* is evidently well adapted for coking coal, and is a great advance in the production of a uniform quality of coke, volatilizing a maximum quantity of sulphur.

The coal is spread evenly over the floor of these ovens, to an average depth of two feet. The heat of the ovens ignites the charge, and, as the coking progresses, the air is more and more excluded by bricking up the door and luting with clay; 48 to 72 hours are usually required to produce coke.

The 3,579 ovens of the Connellsville coke region are all on the beehive plan, receiving an average charge of 100 bushels of coal, dumped through the opening on the crown of the oven, and yielding according to current tradition, 120 bushels of coke.

The yield in Connellsville region, adopting the above data, is as follows:

100 bushels of coal @ 76 ℔s. per bushel, equals 3.39 gross tons.
120 bushels of coke @ 40 ℔s. per bushel, equals 2.14 " "

Loss, - - - - - - 1.25 " "

Yield of coke, 63 *per cent; loss,* 37 *per cent.* One and six-tenth tons of coal to one ton of coke.

The coke is silvery, cellular and tenacious, possessing great calorific power and is comparatively free from impurities.

The Kemble Coal and Iron Company, use beehive ovens to coke for their furnaces at Riddlesburg, in the Broad Top Coal Region.

The process is substantially the same as at Connellsville, and the coke produced very similar in all respects.

The yield is as follows:

Coal charged into oven, - - - - 3.35 gross tons.
Coke taken out, - - - - - 2.74 " "

Loss equals, - - - - - 1.61 " "

The loss of carbon at Connellsville and Broad Top is as follows:

Connellsville, - - - - - - - 9 per cent.
Broad Top, - - - - - - - 22½ "

The result of coking at Hollidaysburg in Belgian ovens, with Bennington coal, is as follows:

Coal used,	6.86 gross tons.
Coke made,	4.81 " "
Loss,	2.05

Yield of coke, 70 per cent; loss, 30 per cent. 1.42 tons of coal to one ton coke.

As the coal used contained 68 per cent of carbon and 8 per cent of ash, and the coke produced, 70 per cent of the coal used, the loss is 10 per cent of carbon in the operation.

In further considering the relative economy of these typical modes of coking, the value of a gross ton of coal will be taken at $100 at the ovens.

I. Pitts or Mounds.

Hollidaysburg and Bennington, 1.67 tons, @ $100 equals	$1 67
Hollidaysburg and Bennington, labor, wood, &c.,	70
Total, one gross ton of coke,	2 37

II. Beehive Ovens.

Connellsville, 1.60 tons; @ $1,	$1 60
Labor, charging and drawing,	61
Total cost, one gross ton coke,	2 21

Broad Top, Kemble Coal and Iron Company.

Coal, 1.58 tons, @ $1,	$1 58
Coking, &c.,	60
Total,	2 81

III. Belgian Ovens—Cambria Iron Company.

Hollidaysburg, Bennington coal—coal used to one ton coke, 142 tons, @ $1,	$1 42
Labor, supplies, &c.,	45
Total cost of one ton coke,	1 87

The cost of the coal and labor of coking one gross ton of coke, by the three methods just considered, is as below:

I. Pits or mounds,	$2 37
II. Beehive ovens,	2 20
III. Belgian ovens,	1 87

Exhibiting an increasing cost from the Belgian ovens to the open pits or mounds. But in the first comparison of costs, no place has been given to the interest on the investment in preparing the several modes for coking, which is quite an important factor.

Estimated cost of *plant* for the production of 100 tons of coke per day:

I. Pits or Mounds..

Leveling coke yard, fixtures, &c.,	$1,000 00
Interest on investment, at ten per cent per year	100 00
Then, $\frac{30,000 \text{ tons}}{\$100}$ = $\frac{1}{3}$ cent per ton per year,	00$\frac{1}{3}$
Cost of coal and labor of coking,	2 37
Total cost,	2 37$\frac{1}{3}$

II. Beehive Ovens.

Eighty ovens, @ $400,	$32,000 00
Interest on investment, 10 per cent per annum,	3,200 00
Annual repairs and renewals, $10 per oven	800 00
Then, $\frac{\$4,000}{30,000 \text{ tons}}$ = 13$\frac{1}{3}$ cents,	13$\frac{1}{3}$
Cost of coal and labor of coking,	2 20
Total cost,	2 33$\frac{1}{3}$

III. Belgian Ovens.

Sixty-five ovens, @ $800,	$52,000 00
Annual repairs to each oven, $15,	310 00
Engine for pushing coke,	3,000 00
Annual repairs to engine,	50 00
Tracks for engines,	300 00
Interest on investment, $55,000, @ 10 per cent	$5,530 00

Then, $5,530+$310+$50=$\frac{\$5,890}{30,000 \text{ tons}}$=19½ cents per
ton nearly 19½
Cost of coal and labor of coking, . . . 1 87
                                          ─────
Total cost, ... 2 06½
                                          ═════

The tracks and cars necessary to supply coal to pits and ovens have not been estimated in the foregoing calculations, as it is presumed these several costs would be about equal, adding to each class one-fourth cent per ton for this source of expense.

The ultimate cost then of one gross ton of coke produced by these three methods is as follows:

Mounds or pits, ... $2 37¾
Behive ovens, ... 2 33⅜
Belgian ovens, ... 2 06¼
                                          ═════

The best cokes have the cell space to the whole mass, as 33 to 67, or as 1 to 2 nearly. But this proportion can differ widely in cokes, giving equally good results in furnace use; 38 to 62 is obtained from a coke of first class order in strength and purity.

Other conditions being equal, *the size of the coke cells is important*, in giving a first quality of furnace fuel, especially in facilitating its combustion by freely receiving carbonic acid gas, formed lower down in the furnace, thus accelerating its solution and yielding abundant carbonic oxide for the deoxidation of the iron ore.

A very dense coke, with diminutive cells, or rather pores, is always undesirable in furnace operations, as it resists solution with an obstinacy that is truly surprising.

A piece of such coke was handed to me which had passed down and out of a fifty-feet furnace, apparently little wasted by its fiery journey. In this connection it may be claimed that anthracite coal is much more dense than coke of any grade, and as the former can be used in the blast furnace, why not the latter?

To this it may be replied, that the operations of these fuels in combustion in a furnace are widely different—the anthracite coal decrepitating, and thus becoming divided into quite small pieces, affording enlarged surface space for solution, whilst, on the other hand, coke is not split into pieces as it approaches the hot zone, and its free combustion is attained only from its large cell structure.

Evidently Mr. I. Lowthian Bell experienced this when he wrote, "My firm has tried these plans" (Belgian ovens), "but found the useful effect in the furnaces inferior to that obtained from coke made in the ordinary oven" (Beehive). "In consequence of this, all the more recently erected ovens have been constructed upon the old fashion."—[I. L. Bell on iron smelting, page 315.]

As the physical structure of coke, and its purity, are the two prime elements which constitute its value as a furnace fuel, it is evident that coke ovens should be planned to satisfy these essential requirements.

It is believed that both these results can be obtained, so far as the qualities of each class of coal will permit, *by shallow chargee in the coke oven.*

A closing consideration in the production of coke claims earnest attention—the means of quenching it in the three methods of its manufacture. The amount of water retained in cokes varies from $\frac{1}{2}$ of 1 per cent to 12 per cent or more, depending on the conditions in which it is quenched.

As has been stated, in the means employed in the pits or mounds, in smothering the coke out with fine dust, only using a very small quantity of water as the last act of the operation, thus giving a very dry coke; with care, certainly a minimum. This is a *very decided advantage in pit coking*, which will be considered hereafter, especially when done near the furnace, giving from the pit a dry fuel.

The coke made in Beehive Ovens is quenched by discharging water into the oven by a hose. The water is quickly converted into steam, which permeates the whole mass of coke, resulting in doing the work with the smallest volume of water and vapor, giving a very dry coke.

The Belgian Oven class is open to serious objection in regard to the manner in which the coke is quenched.

The pushing engine discharges the contents of the oven of red hot coke in a mound 20 feet long, 2 to 3 feet wide and 3 to 4 feet high. A hose is turned on this incandescent mass until it is soaked, pools of water are made on the platform, the vapor escapes, and the coke is charged with moisture from 2 to 12 per cent. It is possible to reduce the average of moisture by a more careful application of the water, but the whole plan of doing this part of the work is essentially clumsy.

As a rule in all methods of quenching coke, the finer the pores or cells of the coke the more moisture it will retain.

(116)

ANALYSIS OF COALS.

Name of Coal.	Fixed Carbon.	Volatile Matter	Ash.	Sulphur.	Water.	Phosphorus.	
Connellsville— Broad Ford	59.616	30.107	8.233	.784	1.210	......	Prof. A. S. McCreath.
Greensburg, Pennsylvania	61.340	33. 50	3.280	0.860	......	......	Prof. A. S. McCreath.
Pittsburg (gas coal)	58.000	36. 00	6.000	.640	......	......	Prof. A. S. McCreath.
Pittsburg, middle bench (near Pittsburgh)	55.608	37.225	4.145	.980	1.040	......	Prof. A. S. McCreath.
Coal Creek, Tennessee	57.520	38. 82	3.090	.200	1.040	......	
Coal Creek, Tennessee	57.690	37. 80	2.550	......	......	......	
Poplar Creek	60.670	36. 53	1.750	0.780	1.750	......	Potter & Riggs.
Poplar Creek	59.470	40. 00	0.530	1.260	......	......	Joliet I. and S. Co.
Careyville (new mine)	56.850	38. 89	3.190	......	1.070	......	
Helenwood	54.240	41. 29	2.640	......	1.830	......	Dr. Peter.
Jellico	60.600	36. 44	1.600	1.160	2.360	......	Regis Chauvenet.
Poplar Creek	56.120	39. 33	2.810	......	1.240	......	Furnished by Mr.H.S.
Crooke Coal and Coke Co.	61.560	34. 53	2.140	0. 88	1.670	.017	Chamberlain, Prest.
Roane Iron Co., (Rockwood	60.110	26. 62	11.520	1. 49	1.750	......	Roane Iron Company
Roane Iron Co.	60.750	32. 59	5.270	......	1.390	......	No analyst given.
Stanley (near Chattanooga)	61.730	26. 70	10.210	.530	1.360	......	Robertson.
Sewanee (Tracy mines.	62.000	25. 41	10.820	......	1.770	......	H. T. Yaryan.
Sewanee " "	63.500	29. 90	6.600	Trace.	......	......	
Soddy mines, (Sewanee seam	64.390	27. 82	6.640	......	1.150	......	Prof. T. E. Wormley.
Emery mines, " "	63.100	27. 70	7.700	.530	.150	......	McCreath & Pohle.
Etna mines, (Kelly seam	74.200	21. 39	2.700	.700	1.300	0.005	Prof. A. S. McCreath.
Bloesburg, Pennsylvania	1.574	21.586	4.753	.907	1.180	......	Prof. W. R. Johnson.
Cumberland, Md	73.500	14. 10	12.400	......	......	......	.B. Britton.
Quinnimont, West Virginia	75.890	18. 19	4. 98	.300	6.740	......	C. E. Dwight.
Longdale	72.320	21.380	5.270	.270	1.030	......	Prof. E. A. Smith.
Pratt mines, Alabama	61.508	31.435	5.416	.918	1.501	......	Eureka I. Co. Chemist.
Helena mines, Alabama	59.580	34.370	6.050	.660	......	......	Otto Wuth.
New Castle, Alabama	59.690	28.240	10.920	.640	0. 50	......	Prof. A. S. McCreath
Blairsville, Pennsylvania	62.220	24. 36	7.590	4.920	.920	......	Prof. A. S. McCreath.
Bennington, Pennsylvania	61.840	27.230	6.930	2.600	1.400	......	Prof. Wormley.
Leetonia, Ohio	56.000	39.600	1.800	.530	2.560	......	Robertson.
Big Muddy, Illinois	59.130	31.930	1.810	.760	6.370	......	
El Moro, Colorado	55.860	38.230	3.590	......	1.320	......	
Pocahontas, Virginia	73.728	20.738	2.984	.618	0.662	0.0013	Prof. A. S. McCreath,

ANALYSIS OF COKES.

Name of Coal.	Fixed Carbon.	Ash.	Sulphur.	Moisture.	Phosphorus.	
Connellsville—Broad Ford	80.576	9.113	0.821	0.630	} 0.011	Prof. A. S. McCreath.
Connellsville—Coketon	80.150	9.650	1.200	......		C. Crowther.
Irvin's gas coal } slack washed. }	88.240	9.477	0.392	1.384	......	Carnegie Brothers.
Bennington	87.580	11.360	1.000	......	......	Prof. A. S. McCreath.
Bloesburg (Arnot)	84.760	13.345	0.908	0.175	......	Prof. A. S. McCreath.
Allegheny River } Lower Freeport.. }	85.777	11.463	2.107	{ 0.330 0.100 }	......	Prof. A. S. McCreath.
Blairsville, Pa	81.450	15.550	1.294	......	0.043	Henry Thomas.
Quinnimont, West Virginia	93.850	5.850	0.300	0.050	......	J. B. Britton.
Longdale, West Virginia	93.000	6.730	0.270	......	......	E. E. Dwight.
Leetonia, Ohio	93.750	5.380	0.870	......	......	Prof. Wormley.
Sewanee seam—Tracy City, Tenn	83.364	15.440	0.142	......	......	W. J. Land.
Etna, (Kelly) Tenn	94.560	4.650	0.790	......	0.008	University of Cincin'ati
Rockwood, Tenn	84.187	14.141	0.182	......	......	W. J. Land.
Dayton, Tenn	84.150	14.880	......	......	......	
Poplar Creek	90.060	5.000	0.570	0.010	0.010	Potter & Riggs.
Poplar Creek	95.240	4.760	......	......	......	Regis Chauvenet.
Pratt, Ala	88.224	11.315	0.563	0.362	......	Prof McCalley.
Helena, Ala	84.025	15.216	0.445	......	......	Prof. McCalley.
Big Muddy, Ill	88.180	10.670	0.610	......	......	T. M. Williamson.
El Moro, Col	87.470	10.680	0.850	......	......	

COKE IN TENNESSEE.

In another part of this work, some extracts have been made from the very able paper of Mr. John Fulton, on the manufacture of Coke. This has been done because there is a general need of information on this important subject, and the book in which it is originally published is accessible to a very small part of the general public.

There is a great need of improvement in the manufacture of Coke in Tennessee. The largest operation in the State is that of the Tennessee Coal, Iron and Railroad Company at Tracy City, and the coal there used is the coking coal of the State from which the largest supply for furnaces must be derived. It is identical in geological position with the coal next to Connellsville the most largely used in the State of Pennsylvania. Reference to the Table of Analysis will show its great resemblance to the Bennington coal. In Mr. Fulton's paper it is shown that Bennington coal, in pits, gave 1 ton of coke to 1.67 of coal; that at Connellsville 1.60 tons of coal make 1 ton of coke; and at Broad Top, 1.58 tons of coal make 1 ton of coke; while under the head of Tracy City mines it is seen that the Superintendent states he is only able to get 105 bushels of coke form 100 bushels of coal, equivalent to 1.90 tons of coal to the ton of coke. Mr. Williams, of the Soddy Coal Company, states that their yield is not over 110 bushels of coke to the 100 of coal. Nor is it probable that much better results are reached at any other coke works.

There is probably no doubt that the best coke yet made in the State was made from the Poplar Creek coal; this coal, however, yields only about sixty per cent of coke, but it contains a minimum amount of ash, hence more of the carbon it contains, its calorific power, is available for action on the iron ore and flux than in those cokes containing a large amount of ash. The highest range of ash in the coke from this coal is five per cent., while it can be seen from the table that Connellsville has 9 to 11, Bennington 11, Blossburg 13, Sewanee 15, Pratt 11, Rockwood 14. It is plain, therefore, that the less of the calorific power of Poplar Creek cokes has to be expended in smelting its own foreign matter than in any of the other cokes, the ash constituents being almost invariably very largely silicious; the percentage of that material in ash of Connellsville coke being 7.210; while in Blairsville coke, from washed coal, with 15.550 of ash, the silica is 9.450. The same proportion would give about three per cent of silica in the Poplar Creek coal. Hence Poplar Creek coal has one per cent. more of heat power than Connellsville, and from four to six per cent. less foreign matter to melt. But it is impossible to get from Poplar Creek more than sixty per cent. of coke, while Connellsville gives sixty-three per cent. Immediately in the neighborhood of the Poplar Creek coal, and also accessible to it by many miles of railroad, is the Sewanee seam in the pitched strata of Walden's Ridge; which by Prof. Wormley's analysis has 63.10 of carbon, 27.70 of volatile matter and 7.70 of ash. There can be no doubt that a mixture of these coals would make a coke coming, at least, very near to the perfect standard, and it is a valuable feature of the northeastern coalfield that they are in such close proximity. The richness of one in the inflammable volatile matter supplies the heating power which with the other alone might cause the loss of a part of its solid carbon.

During the latter part of the year 1882, four ovens were erected at Coal Creek to test the value of the coal of the seam there worked for coke-making, and a considerable quantity of coke has been made chiefly from slack, but the experiments cannot be said to have by any means been a perfect test. The coke has been used in Knoxville, mixed with Connellsville and also with Etna, and the founders speak well of it. There is no reason why it should not make at least as good coke as is made at Larimer Station on the Pennsylvania Railroad from the washed slack of the Pennsylvania gas coal, large quantities of which are used in the furnaces at the Bessemer Steel Works of Carnegie Bros. & Co. The table of analysis shows the great resemblance of these coals. The unwashed slack of that coal contains 11.60 per cent. of ash and 1.26 of sulphur, the washed slack only 6.98 of ash and 96 of sulphur.

In 1870 there were two establishments in the State making coke in ovens, and these two had about thirty ovens. These were the Rockwood and Etna mines. Coke had been made at Tracy City but up to that date only in pits on the ground. There are now 1,000 coke ovens in the State, and about 200 more being built.

These are distributed as follows: Crooke Coal and Coke Company 30, Oakdale Iron Company 61, Roane Iron Company 130, Spring City Coal and Coke Company 27, Dayton Coal Company 24, Soddy and Walden's Ridge Coal Companies 150, Coal Creek Mining Company 4, Etna Coal Company 65, Tennessee Coal and Iron Company at Tracy 404, Victoria 96. Of these, all but the last are in active operation, or if stopped, it is only temporary. The Soddy Coal Company are building 75 more, the Tennessee Coal and Iron Company 140 more at Tracy, and it is probable that the Daisy Coal Company will erect 20 or more.

The production of good and cheap coke is a matter of great importance to Tennessee. While the Western Iron Belt affords unsurpassed facilities for the manufacture of charcoal iron, yet the immense quantity of ore in that region offers abundant and cheap supply for the manufacture of iron with coke, but the problem has been, from whence shall the coke come? If this question was satisfactorily solved there is every probability that a first-class iron furnace would be erected in Nashville. The iron ore can now be reached by three different railroads from numerous large beds of an excellent quality, but the nearest coke is Tracy City, 113 miles distant, and at present that Company has not the capacity to produce more than a supply for their own furnaces. Birmingham is 200 miles distant, with a product that does not exceed the home demand. By present railroad routes, Sparta is 137 miles distant, but if a direct line between the two places was constructed the Sewanee seam of coking coal could be reached in 100 miles. It is plain that from such a line the manufacturing interests of Nashville would receive a great impetus, as not only in that distance would good coking coal be reached, but also the best of steam and domestic coal. When a proper test shall have been made of the Coal Creek coal, alone, or combined with seam B (the Sewanee seam) then the city of Knoxville will be found to have a good coking coal in a short distance, and her iron ores are equally convenient. It is evident from these points that at no very distant day the coke industry of the State will be very largely increased.

Mr. Jos. D. Weeks, in his special report on coke for the Census Bureau, for advance sheets, of which I am indebted to the Hon. C. W. Seaton, Superintendent of the Tenth Census, estimates the total cost of one ton of coke at the best arranged works in the Connellsville region, at $1.15; and the total, including interest on investment of real estate at the Cambria Works, near Connellsville, at $1.49 per ton. The great advantage these works have over any in Tennessee, is in the low cost of mining coal, being there from less than one to one and an eighth cents per bushel, while our lowest is two cents.

The great problem to be solved in the Southern iron-making is the production of cheap coke. It is not probable that we can ever compete in cost with Connellsville, where the mining of the coal costs only one cent per bushel, but taking into consideration the higher charges for railroad transportation which the makers there have to pay, it may be possible to place our coke in a central market like Chattanooga, at as low a rate as Connellsville is placed in Pittsburg. It is evident, however, that such cheap coke must come from a field not now at all, or but little developed, and there can be no doubt that the regular seams of the Upper Measures of Poplar Creek, of the upper Crooked Fork, of Coal Creek and that vicinity offer greater probabilities for having coal mined at low rates than any other part of the State. If those coals can be mined at 1½ cents per bushel, in them will be found the solution of the cheap coke problem, and at the same time the standard of quality will be reached.

Mr. Scott, of Oakdale, says: That in a series of experiments 100 bushels of the Poplar Creek coal made from 120 to 125 bushels of coke; that is 8,000 pounds of coal made from 4,800 to 5,000 pounds of coke, and in a carefully measured month's run of the furnace with a poor hot blast, an average of thirty-seven tons per day was made with seventy-five bushels to the ton. Maj. E. Doud, undoubtedly one of the most careful as well as successful furnace men, says that the Poplar Creek coal has no superior anywhere, and that that region must eventually be the Connellsville of the South.

For a better estimate of the possibility of our competing in cost of coke with the Connellsville region, the following extracts are given from Mr. Weeks' report to the Census Bureau.

Mr. Weeks states that there should be at least 200 acres of coal land for every 100 ovens, and states:

200 acres of land at $400 per acre..............................$80,000
100 ovens at.. 40,000

Total..$120,000

Interest on above at 8 per cent..................................$ 9,600
100 ovens use seven acres of coal at $400 per acre............ 2,800

Amount to be first made in Pennsylvania for interest and depleted investment... $12,000

These 100 ovens estimated to make 39,000 tons of coke per annum, and that hence the sum of thirty-two cents per ton is necessary to cover interest and replace capital.

The cost at Frick & Co.'s Valley Works is estimated as—

Coal for one ton of coke...	.38
Drawing coke...	.25
Loading, etc...	.10
Repairs..	.83—.10
Interest, etc., as above..	.32
	1.15

Mr. Weeks very sensibly states that he thinks this too low, and from my own observations at Valley Works, the best arranged plant, and other of Frick & Co.'s coke works, I am satisfied that he is correct; and if it be true, that it is true only of the mines and works at their best arranged place.

At Morrell and Wheeler, in the Connellsville region, the Cambria Iron Company have 500 ovens, which cost about $500 each, and they pay 25 cents per ton for room coal and 32 cents for heading coal. The figures he gives are as follows:

Mining coal per ton (2,000 lbs)....................................	.276
Hauling...	.073
Hoisting and dumping...	.038
Superintendent, Foreman and Clerk............................	.016
Lumber, ties and props..	.029
Repairs and supplies...	.068
Cost of coal per ton at ovens.......................................	.50
1.60 tons at 50 cents..	.80
Labor (draining, loading, charging, Sup't and Clerk)......	.412
Supplies..	.026
Repairs...	.056
Cost of coke per ton..	1.29

At these Works the amount estimated to pay for improvements and interest is 20 cents, hence their cost of one ton of coke is $1.49,

It is estimated that every acre of the Connellsville field will average coal enough to make 5,500 tons of coke, or say 8,800 tons of coal.

Now for comparison, let us estimate for the great area of Upper Measure coal which comprises the upper Crooked Fork, upper New River, the Poplar and Coal Creek region, and some north thereof. The main seam now worked is very regularly 5 to 5½ feet thick at Poplar Creek. Assuming a yield to the acre of only 5,000 tons of coal, at 1.60 to the ton, for making 39,000 tons of coke as stated above, thirteen acres would be annually worked out. A high valuation of any of this land would be $100 per acre, and ovens would not cost over $300 each. Therefore we estimate:

400 acres of coal land at $100 per acre	$40,000
100 ovens	30,000
Total	$70,000
Interest on $70,000	$5,600
13 acres of coal at $100 per acre	1,300
Total	$6,900
Mining coal for one ton of coke at 1½ cts. per bushel, 1.60 tons	.60
Mine expenses	.20
	.80
Charging, drawing, loading and repairs	.45
	$1.25
Interest and loss of Investment per ton	·.16⅔
Total cost of 1 ton of coke from Poplar Creek Coal	$1.41⅔

RECAPITULATION.

Cost of one ton of coke at Frick & Co.'s Valley Works..... $1.15
Cost of one ton of coke at Cambria Company's Connellsville Works...... 1.49
Cost of one ton of coke at Poplar Creek.......... 1.41⅔

In Mr. Fulton's paper it has been seen that he makes the cost of coke much higher, but that result is arrived at chiefly from his placing the coal at $1.00 per ton. He does so simply to institute a comparison between methods of coking. It is a singular commentary on the above figures that coke is sold in the Connellsville region at this date, September 1st, at $1.00 per ton of 2,000 lbs., for furnace coke and $1.25 for foundry coke. The average price for 1882, however, was $1.50 per ton. From the ovens to Pittsburg, a distance of 40 to 50 miles, the freight rate is now $1.00 per ton; from the ovens to Chicago $4.67 per ton. The distance from

Poplar Creek or the upper Crooked Fork to Chattanooga would be from 95 to 100 miles, and at present rates the freight would not be over $1.00 per ton.

Mr. Weeks in his very excellent and thorough compilation, does not give any cost for the manufacture of coke from the washed slack at Irvin and Larimer, which is to be regretted; but coke cannot be profitably made from the washed slack at Coal Creek, because the land owners will not make the royalty on it less than one-half a cent, at which rate, and any price on it, its cost would reach the amount at which the run of the mines can be produced on Crooked Fork and Poplar Creek. The Crooked Fork region is immediately northwest of Poplar Creek, on the opposite side of the mountains, and has the same coal seams as those known and opened on the waters of Poplar Creek. It has not now any connection by rail with the Cincinnati Southern, but such will undoubtedly be made at some not far distant time.

In a conversation with Prof. M. M. Duncan, Chemist and Superintendent of the Roane Iron Company's furnaces; who is a gentleman of great energy and high scientific attainments, the writer was lately informed that in experiments made in their furnaces Prof. Duncan had succeeded in making an excellent pig for steel manufacture from the red fossil ores by the Basic process, and that there was only one difficulty, the large amount of ash in their coke. The mixture with Poplar Creek coal, spoken of on page 119, was alluded to, and he stated that such a mixture would undoubtedly be of great advantage. No one can study the details of our furnace workings and not be satisfied that however cheap our ores, there must be a decided improvement in the quality of our coke if we propose to successfully compete with the ironmakers of the Pittsburg region. The fact is now plain that our chief coaking coal—the Sewanee seam—does not make at best over 110 bushels of coke to the 100 of coal, that it contains from 14 to 18 per cent. of ash, and that from 100 to 105 bushels of coke to the ton of pig is the usual average, and even this inferior coke is sold to our furnaces at full 50 per cent more that the best Connellsville costs the furnaces in Pittsburg.

The census statistics of Mr. Weeks, show that our percentage yield is only 51, being 12 less than Pennsylvania, and 5 less than Ohio, yet Tennessee is he fourth State in the amount of coke produced, using more coal, but making less coke than West Virginia.

The average selling price of coke in 1880 for the whole United States was $1.95, while for Tennessee it was $2.32 per ton, and Alabama was up to $3.52 per ton, Georgia's one mine sold coke at $2.00, and Pennsylvania's many at an average of $1.81 per ton. The same statistics place the average cost of Tennessee coke at $1.86 per ton against $1.39 in Pennsylvania, $1.98 in Georgia and $1.96 in West Virginia. Whatever may be the positive accuracy of these figures, they at least furnish points for the consideration of our coke manufacturers. If we expect to make coke to compete with Northern manufacturers, we must firstly, let third-class coals alone; and secondly, improve our mode of manipulating those we have of the first and second class.

COAL CONSUMPTION AND PRICES.

The consumption as well as production of coal in the State of Tennessee is chiefly a growth of the past seventeen years, and in a large measure of the last thirteen. Previous to 1860, stone coal was very little used outside the cities of Memphis and Nashville, except for blacksmith purposes. Previous to 1860 the coal supply of Nashville was chiefly derived from the Cumberland River, a little coming from Sewanee, and the gas company getting their coal from Pittsburg. Memphis drew her supply from Pittsburg, and Knoxville hauled from Poplar Creek, while the consumption in Chattanooga was a mere trifle. Even as late as 1870, Nashville consumed only about 40,000 tons. During the year 1882, the Louisville & Nashville Railroad reports that there was brought to the city in their cars from the Kentucky mines 68,266 tons of coal, and from Alabama 17,553 tons, while the Nashville & Chattanooga Railroad brought from the Tennessee mines about 30,000 tons. The city of Memphis in 1882 received 1,000,000 bushels (40,000 tons) of coal by rail, probably all from Kentucky mines, and 1,425,000 bushels (57,000 tons) by river, probably all from Pittsburg.

The entire shipments over the Nashville & Chattanooga Railroad for the same time from all points were:

	COAL		COKE	
	Bushels North.	Bushels South.	Bushels North.	Bushels South.
Cowan	809,905	1,167,900	35,468	1,964,454
Shellmound	2,700	1,105,561	....	1,723,018
Whitesides	87,701	131,223	195,340	119,194
Chattanooga	10,320	....	....	....
Total	960,663	2,404,584	250,922	3,906,166

For the same year the Ohio branch of the East Tennessee, Virginia and Georgia Railroad transported 199,500 tons from Coal Creek, of which about 50,000 tons was for the use of the railroad, and 48,391 tons for Knoxville, 19,644 tons for stations on their line, and 80,801 for points beyond.

The Cincinnati Southern has done a very heavy business, which is constantly increasing, but the road was in the hands of its present managers only a part of 1882, and Mr. E. P. Wilson, the General Freight Agent, writes that it is impossible to give an accurate statement of amounts and places of destination.

The building of new railroad lines and the consolidation of old ones under one management whereby freights are cheapened and the shipments simplified, has already opened new markets for Tennessee coal and coke and is destined to still further extend the area which they will supply. The increase in the last two years is remarkable. In 1880 there were twenty mines, employing 1,002 persons, paying out $386,765 in wages, and producing coal valued at $638,954. In 1882 there were thirty-five mines, employing 2,481 persons, paying out $619,263 in wages, and producing coal valued at $1,106,737. In 1880 there was used in the State 74,408 tons of coke valued at $182,241, in 1882 one company alone produced coke valued at $180,300.

The prices of coal in Nashville are: For steam coal (slack) 7½ cents per bushel; slack and nut mixed, 9 cents; nut alone, 11 cents; run of mines, 11 cents; lump, 13 cents; each delivered, but for some on cars the prices are two cents per bushel less. Coal for domestic use sells in summer from $2.50 to $3.00 for seventeen bushels delivered; in winter from $2.75 to $3.25 for seventeen bushel lots, depending on the size of the coal.

In Knoxville steam coal on cars sells at $2.50 to $2.25 per ton of 2,000 pounds; domestic coal delivered at $4.00 per ton.

In Chattanooga steam coal $1.75 to $2.00 per ton; domestic coal $2.00 to $2.50 delivered.

INDEX.

	PAGE.
Anderson County Coal Company	45
Analysis of Coals	116
Analysis of Cokes	117
Area of Coal-field	56
Black Diamond Coal Company	44
Bon Air Section	12
Byrd & Denning Mines	58
Carey Lands	36
Careyville District	33
Campbell County Coal Company	34
Central Coal Company	46
Coal Creek Mining Company	46
Coal Creek District	37
Coal Creek Shipments	41
Chattanooga Division	48
Crooke Coal and Coke Company	54
Coal Identity	25
Coke in Tennessee	118
Coal Consumption and Prices	122
Dayton Coal Company	64
Daisy Coal Company	70
Elevations	6 to 12
Emory Mines	58
Elk Fork District	32
East Tennessee Coal Company	33
Elk Gap	35
Etna Mines	72
Eureka Coal Company	52
Horrizontal Strata	68
Heck's Mines	47
Helenwood	55
Henry H. Wiley	52
Inclined Strata	56
Iron Ores of Tennessee	95
Jellico Coal Company	33
Knoxville Iron Company's Mine	42
Knoxville Division	41
Lower Coal Measures	87

	PAGE.
Mt. Carbon Coal Company	52
Markets and Freights	47
Oakdale Company	52
Oliver Coal Company	52
Poplar Creek District	48
Prices of Coal	123
Plateau Dirtrict	53
Report of Mines	29
Roane Iron Company's Mines—Rockwood	59
Sand Mountain	6
Sequatchie Valley	84
Southern Coal Company	52
Standard Coal Company	32
Sharp's Coal	36
Star Coal Company	44
Spring City Coal Company	61
Sewanee District	76
Soddy Coal Company	68
Section at Tracy City	11
Section Across the State	17
Section at Coal Creek	19
Section at Little Emory	18
Section of Salt Wells at Winter's Gap—Coal Creek	21
Section in Pennsylvania	24
Section on the Sequatchie Valley & Tennessee River Railroad	63
Section at Soddy	66
Section at Daisy	67
Section at Etna	68
Topography	6
Tracy City Mines	78
Tennessee Coal, Iron and Railroad Company	78
Tennessee River and Sequatchie Valley Railroad	58
Victoria Mines	86
Walden's Ridge District	56
Walden's Ridge Coal Company	70
White's Creek	61
Water Transportation	53

www.ingramcontent.com/pod-product-compliance
Lightning Source LLC
Chambersburg PA
CBHW020111170426
43199CB00009B/487